Erwin Dee Kord (Ed.)

Illinois Route 105

Erwin Dee Kord (Ed.)

Illinois Route 105

Illinois Route 48, Interstate 72, Macon County, Piatt County, Illinois Department of Transportation

Solv

Imprint

Permission is granted to copy, distribute and/or modify this document under the terms of the GNU Free Documentation License, Version 1.2 or any later version published by the Free Software Foundation; with no Invariant Sections, with the Front-Cover Texts, and with the Back- Cover Texts. A copy of the license is included in the section entitled "GNU Free Documentation License".

All parts of this book are extracted from Wikipedia, the free encyclopedia (www.wikipedia.org).

You can get detailed informations about the authors of this collection of articles at the end of this book. The editors (Ed.) of this book are no authors. They have not modified or extended the original texts.

Pictures published in this book can be under different licences than the GNU Free Documentation License. You can get detailed informations about the authors and licences of pictures at the end of this book.

The content of this book was generated collaboratively by volunteers. Please be advised that nothing found here has necessarily been reviewed by people with the expertise required to provide you with complete, accurate or reliable information. Some information in this book maybe misleading or wrong. The Publisher does not guarantee the validity of the information found here. If you need specific advice (f.e. in fields of medical, legal, financial, or risk management questions) please contact a professional who is licensed or knowledgeable in that area.

Any brand names and product names mentioned in this book are subject to trademark, brand or patent protection and are trademarks or registered trademarks of their respective holders. The use of brand names, product names, common names, trade names, product descriptions etc. even without a particular marking in this works is in no way to be construed to mean that such names may be regarded as unrestricted in respect of trademark and brand protection legislation and could thus be used by anyone.

Cover image: www.ingimage.com
Concerning the licence of the cover image please contact ingimage.

Publisher:
Solv is a trademark of
International Book Market Service Ltd., 17 Rue Meldrum, Beau Bassin, 1713-01 Mauritius
Email: info@bookmarketservice.com
Website: www.bookmarketservice.com

Published in 2011

Printed in: U.S.A., U.K., Germany. This book was not produced in Mauritius.

ISBN: 978-613-8-84508-9

Contents

Illinois_Route_105   1

Illinois_Department_of_Transportation   2

Illinois_Route_48   6

Interstate_72   7

Bement,_Illinois   11

Illinois_Route_104   14

Illinois_Route_47   15

Macon_County,_Illinois   18

Piatt_County,_Illinois   22

Bryant_Cottage_State_Historic_Site   28

Illinois   29

References

Article Sources and Contributors   55

Image Sources, Licenses and Contributors   57

Illinois_Route_105

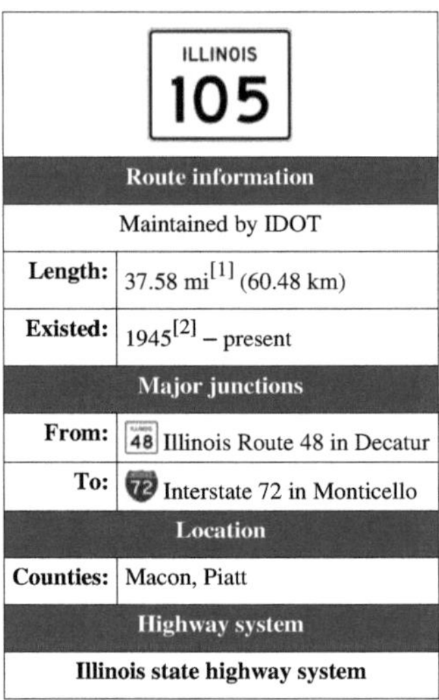

Illinois Route 105 (**IL-105**) is a highway in the U.S. state of Illinois. It is an east–west highway that runs from Illinois Route 48 in Decatur to Interstate 72 near Monticello. Illinois 105 is 37.58 miles (60.48 km) long.[1]

Route description

The state highway serves a corn-and-soybean-growing region in Macon County and Piatt County. Roads such as IL-105 carry corn and soybeans to the refineries of Decatur for the manufacture of commodities such as soybean oil and corn syrup.

Rural towns along Illinois Route 105 include Bement and Cerro Gordo. The highway also runs past the Bryant Cottage State Historic Site.

History

SBI Route 105 was what Illinois Route 104 is now from Meredosia to Quincy from 1924 thru 1937. In 1945, the number was reused on former Illinois Route 47 north of Decatur. This is the current routing for Illinois 105.[2]

External links

- Illinois Highway Ends: Illinois Route 105 [3]

References

[1] Illinois Technology Transfer Center (2006). "T2 GIS Data" (http://www.dot.state.il.us/gist2/select.html). . Retrieved 2007-11-08.

[2] Carlson, Rich. Illinois Highways Page: Routes 101 thru 120 (http://www.n9jig.com/101-120.html). Last updated March 15, 2005. Retrieved April 28, 2006.

Illinois_Department_of_Transportation

Illinois Department of Transportation (IDOT)	
Illinois Department of Transportation	
Agency overview	
Formed	1972
Preceding agency	Illinois Department of Public Works and Buildings
Jurisdiction	Illinois
Headquarters	2300 S. Dirksen Parkway, Springfield, Illinois
Agency executive	Ann L. Schneider[1] , Acting Secretary of Transportation
Parent agency	State of Illinois
Website	
http://www.dot.state.il.us	

The **Illinois Department of Transportation (IDOT)** is a state agency in charge of state-maintained public roadways of the U.S. state of Illinois. In addition, IDOT provides funding for rail, public transit and airport projects and administers fuel tax and federal funding to local juridictions in the state. The Secretary of Transportation reports to the Governor of Illinois. IDOT is headquartered in unincorporated Sangamon County, located near the state capital, Springfield. In addition, the IDOT Division of Highways has offices in nine locations throughout the state.[2] [3]

The mission of IDOT is to provide safe, cost-effective transportation for Illinois in ways that enhance quality of life, promote economic prosperity and demonstrate respect for the environment.

Organization

As of February 2009, the Illinois Department of Transportation was divided into the following offices and divisions:

Offices

- The **Office of Business and Workforce Diversity** oversees the implementation of directives, policies and strategies for departmental business diversity efforts.

- The **Office of Chief Counsel** provides legal counsel to the Department on policy issues and proposed actions affecting any of its operating divisions or staff offices. The Office is responsible for the prosecution and defense of all litigation involving the Department in cooperation with the Illinois Attorney General. The Office of Chief Counsel administers tort liability claims, property damage claims and uncollectable receivables as well as processes lien and bond claims against contractors. The Office coordinates the purchase and service of all insurance policies and administers the Department's self-insurance program.

- The **Office of Finance and Administration** develops and administers the Department's budget; manages the Department's personnel system; provides accounting and auditing functions; provides centralized business services functions and facilities management; and provides management information capabilities.

- The **Office of Communications** was created in 2009 by combining the Office of Governmental Affairs and the Office of External Affairs. The Office of Communications develops and implements the Department's public affairs policies, plans and programs. This includes developing the Department's policy goals and positions;

prepares and implements state legislative programs and strategies; analyzes issues of special interest to the Secretary; and represents the Secretary before various state and national organizations. It's primary objectives are to ensure adequate information toward increasing public involvement in the transportation planning process; assist the news media in the coverage of agency activities; increase the Department's sensitivity to its public and interpret public opinion so that agency programs and regulations will be realistic and acceptable; and to mobilize support for the Department and its programs.

- The **Office of Planning and Programming** develops programs to improve the state transportation system. This includes working with metropolitan planning organizations in ten of the state's urbanized areas to develop programs relating to urban transportation; coordinating a surveillance program to monitor the physical condition of the transportation system, the level of service provided and the need for improvement; evaluating proposals for major investments in the transportation system and overall benefits to be gained. The Office ensures the continuation of state rail services where the potential for efficiency and economy are most favorable and minimizing the expenditure of public funds for rail subsidies. The Office works closely with the Chicago Metropolitan Agency for Planning (CMAP), which serves as a forum for transportation decision making by local elected officials in northeastern Illinois. The Office develops and implements Federal legislative initiatives as well as the initiation and coordination of policy statement and papers which serve as guides for departmental actions on a broad spectrum of transportation issues.

- The **Office of Quality Compliance and Review** independently tests the Department's internal control systems to further ensure to the Secretary and to the public the adequacy of the policies, regulations and procedures and to recommend improvements.

Divisions

- The **Division of Aeronautics** coordinates and implements programs concerning air safety, airport construction and other aeronautical related issues in Illinois.

- The **Division of Highways** develops, maintains and operates the state highway system. The central bureaus of the Division developing policies, procedures, standards and guidelines to accomplish the Department's highway system improvement objectives. The central bureaus monitor the nine district programs to ensure statewide uniformity of policy interpretation and compliance and to ensure program coordination with federal, state and local agencies.

- **District Offices**
  - District 1 - Schaumburg, Illinois (covers Chicago Metropolitan Area)
  - District 2 - Dixon, Illinois
  - District 3 - Ottawa, Illinois
  - District 4 - Peoria, Illinois
  - District 5 - Paris, Illinois
  - District 6 - Springfield, Illinois
  - District 7 - Effingham, Illinois
  - District 8 - Collinsville, Illinois (covers St. Louis Metropolitan Area)
  - District 9 - Carbondale, Illinois
- The **Division of Public and Intermodal Transportation** promots and assures safe and efficient mass transportation systems and services in the State of Illinois by developing and recommending policies and programs; developing, implementing and administering operating, capital and technical program projects and grants; and coordinating and participating in local and statewide planning and programming activities.
- The **Division of Traffic Safety** providing Illinois motorists, cyclists and pedestrians with a safe environment by promoting the reduction of traffic fatalities, injuries and accidents. The Division develops and promulgates regulations in areas of accident reporting, hazardous materials transportation, vehicle inspection, safety responsibility, cycle rider training and highway safety Federal Section 402, 408 and 410 Grants.

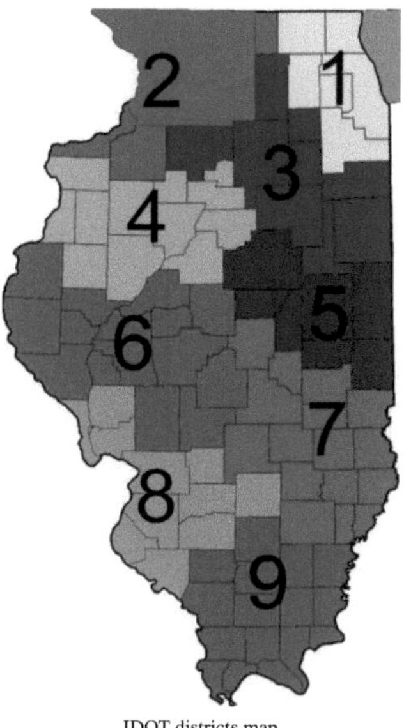

IDOT districts map

History

The Illinois Department of Transportation was created by the 77th Illinois General Assembly on January 1972. The department absorbed the functions of the former Department of Public Works and Buildings, acquired some planning and safety inspection functions of other state agencies, and received responsibility for state assistance to local mass transportation agencies such as te Chicago-area Regional Transportation Authority, which was in the process of being formed at this time. The Division of Aeronautics was added in 1973.

On June 18, 2005 IDOT became the first state transportation agency to achieve ISO 9001:2000 certification for 23 key processes located in the Central Administrative Office and regional District Six. On July 6, 2006 that certification was expanded to encompass all processes involved in the planning, design, and construction of road and bridge improvements, maintenance of roads and bridges, and administrative oversight in the Central Administrative Office and District Six. An official with IDOT, John D. Baranzelli, wrote a book called *Making Government Great Again* on the transition to ISO9001:2008 standards in July 2009 for the American Society for Quality publication arm (ASQ Press).[4]

Notes and references

External links

- Illinois Department of Transportation (http://www.dot.state.il.us/)

Illinois_Route_48

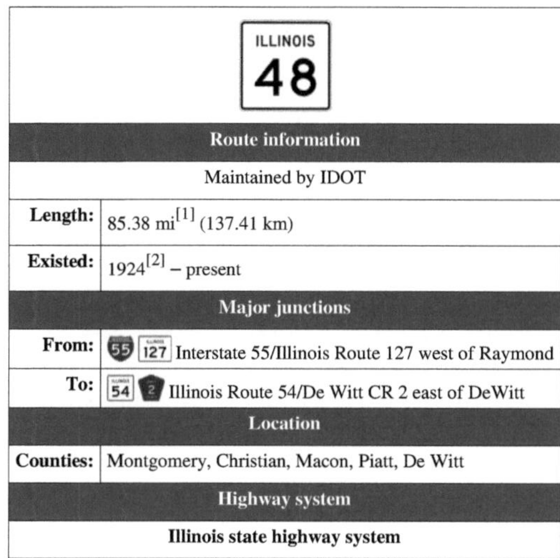

Route information	
Maintained by IDOT	
Length:	85.38 mi[1] (137.41 km)
Existed:	1924[2] – present
Major junctions	
From:	Interstate 55/Illinois Route 127 west of Raymond
To:	Illinois Route 54/De Witt CR 2 east of DeWitt
Location	
Counties:	Montgomery, Christian, Macon, Piatt, De Witt
Highway system	
Illinois state highway system	

Illinois Route 48 is a north–south[3] highway with its southern terminus at I-55 and Illinois Route 127 in Raymond and its northern terminus at Illinois Route 54 east of Clinton. This is a distance of 85.38 miles (137.41 km).[1]

Route description

Illinois 48 runs southwest to northeast from Interstate 55 to east of Clinton. It is an undivided surface street for its entire length. It also serves the cities of Raymond, Harvel, Taylorville and Decatur.

History

SBI Route 48 ran from Onarga to Raymond via the current Illinois 48 and Illinois 54. The portion from DeWitt to Onarga became U.S. Route 54 in 1942, but in 1972 U.S. 54 was pulled back to its current terminus in Pittsfield, and Illinois 54 was implemented from Springfield to Onarga.

References

[1] Illinois Technology Transfer Center (2006). "T2 GIS Data" (http://www.dot.state.il.us/gist2/select.html). . Retrieved 2007-11-08.

[2] Carlson, Rich. Illinois Highways Page: Routes 41 thru 60 (http://www.n9jig.com/41-60.html). Last updated March 15, 2005. Retrieved April 7, 2006.

[3] Illinois Highways Ends. Route 48 (http://highwayexplorer.com/il_EndsPage.php?id=1048§ion=1). Retrieved April 7, 2006.

Interstate_72

Interstate 72Route informationLength: 179.29 miFederal Highway Administration (2002-10-31). "FHWA Route Log and Finder List: Table 1". . Retrieved 2007-03-28. (288.54 km)Major junctionsWest end: U.S. Route 36 in MissouriUS 36 / U.S. Route 36 Business (Hannibal, Missouri)US 36 Bus. / U.S. Route 61 in MissouriUS 61 in Hannibal, MissouriHannibal, MO Interstate 172 (Illinois)I-172 in Fall Creek, IllinoisFall Creek, IL U.S. Route 54 in IllinoisUS 54 in Griggsville, IllinoisGriggsville, IL U.S. Route 67 in IllinoisUS 67 in Jacksonville, IllinoisJacksonville, IL Interstate 55 (Illinois)I-55 in Springfield, IllinoisSpringfield, IL U.S. Route 36 in IllinoisUS 36 / U.S. Route 51 in IllinoisUS 51 in Decatur, IllinoisDecatur, IL U.S. Route 51 Business (Decatur, Illinois)US 51 Bus. in Decatur, IllinoisDecatur, IL Interstate 57 (Illinois)I-57 in Champaign, IllinoisChampaign, ILEast end: Country Fair Drive in Champaign, IllinoisChampaign, ILHighway systemMain route of the Interstate Highway System Interstate 72 (I-72) is an Interstate Highway in the midwestern United States. Its western terminus is in Hannibal, Missouri, at an intersection with U.S. Route 61; its eastern terminus is at Country Fair Drive in Champaign, Illinois. In 2006, the Illinois General Assembly dedicated all of Interstate 72 as Purple Heart Memorial Highway. The stretch between Springfield and Decatur, IllinoisDecatur is also called Penny Severns Memorial Expressway, and the section between Mile 35 and the Mississippi River is known as the Free Frank McWorter Historic Highway.Route description LengthsmilemikilometerkmMissouriMO2 3 IllinoisIL182 295 Total 184 298 I-72 runs for just over 2 miles (3.2 km) in the state of Missouri. Its western terminus is an interchange with U.S. Route 61 to the Mark Twain Memorial Bridge over the Mississippi River. This bridge connects the city of Hannibal, MissouriHannibal with Illinois. Presently, there are only two exits for I-72 in Missouri. In Illinois, I-72 runs for 182 miles (293 km). The portion of I-72 and I-172 from Springfield to Quincy, IllinoisQuincy is commonly referred to as the Central Illinois Expressway (CIE). As of 2007, I-72 has one business route — Interstate 72 Business (Jacksonville, IL)Business Loop 72 in Jacksonville, IllinoisJacksonville. A section of I-72, north of Seymour, Illinois, facing west. Near Valley City, IllinoisValley City at mile marker 42 are the Valley City Eagle Bridges. These two individual two-lane spans bridge the Illinois River in rural west-central Illinois. Near mile marker 78, a sign marks 90th meridian west90 degrees longitude. At its eastern terminus in Champaign, IllinoisChampaign, I-72 continues as Church Street (westbound) and University Avenue (eastbound), which stay as one-way streets for an additional 3 miles (4.8 km) into downtown Champaign. History Until the mid-1990s, I-72 ran from Springfield, IllinoisSpringfield at Interstate 55 to Champaign, IllinoisChampaign at Interstate 57. On June 9, 1991, the American Association of State Highway and Transportation Officials (AASHTO) approved the establishment of Interstate 172 from the western terminus of I-72 at Springfield to Fall Creek, IllinoisFall Creek, 4 miles (6.4 km) east of Hannibal, Missouri, though it was contingent on Federal Highway Administration (FHWA) approval. The Federal Highway AdministrationFHWA preferred to designate the route I-72.American Association of State Highway and Transportation Officials (1991-06-09). "Report of the Special Committee on U.S. Route Numbering to the Executive Committee". . Retrieved 2007-12-19. "Interstate 172 Illinois". Interstate-Guide.com. .After discussions regarding extending an Interstate-standard highway through the state of Missouri, on April 22, 1995, American Association of State Highway and Transportation OfficialsAASHTO approved another renumbering. I-172 was renumbered in its entirety as I-72. The U.S. 36 extension west of Fall Creek was also given the I-72 designation. The Illinois Route 336 expressway was renumbered to I-172 from Fall Creek to Fowler, IllinoisFowler.American Association of State

Highway and Transportation Officials (1995-04-23). "Report of the Special Committee on U.S. Route Numbering to the Standing Committee on Highways". . Retrieved 2007-12-19.Prior to September 2000, Mark Twain Avenue (old US 36) was composed of the current Mark Twain Avenue (now Route 79) and the portion of I-72 and US 36 west of Exit 157 to the Hannibal city limits. Route 79 terminated at the foot of the old Mark Twain Memorial Bridge at the corner of Third Street and Mark Twain Avenue. Signs along the four-lane expressway portion of Mark Twain Avenue marked the route as "Future I-72", while signs along what is now Route 79 had I-72 trailblazers to direct drivers to the temporary terminus at Fall Creek, Illinois. When the new Mark Twain Memorial Bridge was completed in September 2000, I-72 was routed over the new bridge, along with US 36. Route 79 was extended along Mark Twain Avenue to terminate at Exit 157. The portion of I-72 and US 36 west of Exit 157 is now referred to as the V.F.W. Memorial Highway. Originally, I-72 opened with a posted speed limit of 65 mph (105 km/h). However, when it first opened, some drivers were confused and were driving it at 35 mph (56 km/h), the old posted speed limit on Mark Twain Avenue. Later, Hannibal convinced MoDOT to lower the speed limit along I-72 and US 36 within the Hannibal city limits to 55 mph (89 km/h). Chicago-Kansas City expressway The concept of I-72 across Missouri was to create the Chicago – Kansas City Expressway, a rural four-lane highway across northern Missouri and west central Illinois from Cameron, Missouri at I-35 to Springfield, Illinois at Interstate 55 (Missouri)I-55. This would provide a series of rural 4-lane highways (I-35, US 36, I-72, and I-55) connecting Chicago to the North American Free Trade Agreement (NAFTA) Corridor (High Priority Corridor 23). This would reduce the amount of through traffic, primarily truck traffic, in the St. Louis, MissouriSt. Louis, Des Moines, IowaDes Moines, and Quad Cities metropolitan areas by serving as an alternate route for Interstate 70 (Missouri)I-70 and Interstate 80 (Iowa)I-80. The Missouri portion of this route is designated as part of High Priority Corridor 61. Based on the 157-mile (253 km) marker at Route 79, when U.S. Route 36 (Missouri)US 36 is upgraded to Interstate standards across Missouri, the future western terminus of I-72 would be at Cameron, Missouri at the intersection with Interstate 35 (Missouri)I-35. Currently, the west end of I-72 route is west of U.S. Route 61 (Missouri)US 61 and flows concurrent with US 36 into Illinois. In 2004, US 36 was upgraded to a 4-lane expressway between US 61 and U.S. Route 24 (Missouri)US 24 at the Rocket Junction (7 miles). There are three exits along this expressway: Veterans Road, Shinn Lane (Hannibal Regional Hospital), and U.S. Route 24 (Missouri)US 24 East (Future Hannibal Bypass) at the Rocket Junction. This expressway is up to interstate standards (completed August 2007). Also, an interchange with Route 15 (Missouri)Route 15 was installed in Shelbina, MissouriShelbina. At Clarence, MissouriClarence, US 36 resumes 4-lane expressway status. Due to funding priorities, upgrading US 36 between Macon, MissouriMacon and Hannibal, MissouriHannibal was a low-priority project and was officially tabled by MoDOT. MoDOT committed to constructing the four-lane highway as a non-interstate expressway only if the five counties served by US 36 east of Macon would contribute half of the $100 million cost. Four-lane construction The first proposition of the expressway construction, Proposition 36, was passed by a majority of the voters in Macon County, MissouriMacon, Marion County, MissouriMarion, Monroe County, MissouriMonroe, and Shelby County, MissouriShelby counties. However, residents of Ralls County, MissouriRalls County rejected the proposition, citing lack of economic benefit for the county. US 36 cuts through the very northwest corner of Ralls County from Monroe City, MissouriMonroe City to the BNSF viaduct, a distance of 4 miles (6.4 km). Since the proposition failed in Ralls County, the entire proposition failed. Businesses and voters in the other four counties still strongly supported the four-lane expressway project. At the next election, on August 3, 2005, voters of all five counties approved Proposition 36B, which excludes Ralls County from the Transportation Development District and allows for the construction of a 4-lane US 36 to be constructed without Ralls County's participation. The Proposition passed 66 percent to 34 percent and passed by a majority in all 5 counties. In March 2008, the revised total cost of the project was estimated at $89 million.Missouri Department of Transportation, Northeast District (2007), Route 36 four-lane, MoDOT Additionally: A fifteen year, 1/2-cent sales tax was levied in Macon, Marion, Monroe, and Shelby counties for their portion (42%) of the construction costs for the project. The Ralls County portion was covered by a $7 million earmark for US 36 from Congressman Kenny Hulshof. The remainder of the project costs, including construction, engineering, right of way acquisition, survey, etc. were funded by MoDOT. Road construction to complete the 52.4

miles (84.3 km) between Hannibal and Macon began in 2007. According to MoDOT, the estimated completion date of four lanes from Hannibal, MissouriHannibal to Monroe City, MissouriMonroe City, 11.5 mi (18.5 km), was September 2008, from Monroe City to Shelbina, MissouriShelbina, 18.4 mi (29.6 km), was December 2009, and from Shelbina to Macon, MissouriMacon, 22.5 mi (36.2 km), was December 2010. In August 2010, the 4-lane expressway was completed from Macon to Hannibal, completing Missouri's portion of the Chicago-Kansas City Expressway.Missouri Department of Transportation, North Central District (2007), U.S. 36 4-Laning Project: Contract Awarded for Route 36 Four-lane Project from Shelbina to Macon, MoDOTFuture Based on the 157-mile (253 km) marker at Missouri 79, when US 36 is upgraded to Interstate standards across Missouri, the future western terminus of I-72 would be at Cameron, Missouri at the intersection with I-35. The concept of I-72 across Missouri was to create the Chicago – Kansas City Expressway, a rural 4-lane highway across northern Missouri and west central Illinois from Cameron, Missouri at I-35 to Springfield, Illinois at I-55. This would provide a series of rural 4-lane highways (I-35, US 36, I-72, and I-55) connecting Chicago to the North American Free Trade Agreement (NAFTA) Corridor (High Priority Corridor 23). This would reduce the amount of through traffic, primarily truck traffic, in the St. Louis, MissouriSt. Louis, Des Moines, IowaDes Moines, and Quad Cities metropolitan areas by serving as an alternate route for I-70 and I-80. The Missouri portion of this route is designated as part of High Priority Corridor 61. Due to funding priorities, it was initially determined that upgrading US 36 between Macon, MissouriMacon and Hannibal, MissouriHannibal was a low-priority project and was officially tabled by MoDOT. MoDOT committed to building the four-lane highway as a non-interstate expressway only if the five counties served by US 36 east of Macon would contribute half of the $100 million cost. The upgrade to 4-lane expressway on US 36 has been completed in the fall of 2010 and the route has been marked with CKC signs from Hannibal, MO to Cameron, MO. Exit list State County Location Mile Exit Destinations Notes MissouriMarion_County,_MissouriMarionHannibal,_MissouriHannibal U.S. Route 36 in MissouriUS 36 west – Monroe City, MissouriMonroe CityContinuation beyond US 36 Bus. / US 610.0– U.S. Route 36 Business (Hannibal, Missouri)US 36 Bus. east / U.S. Route 61 in MissouriUS 61 (McMasters Avenue) – New London, MissouriNew London, Palmyra, MissouriPalmyraWest end of US 61 Bus. overlap1.0157 U.S. Route 36 Business (Hannibal, Missouri)US 36 Bus. west / U.S. Route 61 Business (Hannibal, Missouri)US 61 Bus. south / Missouri Route 79Route 79 south / Route N north – Downtown Hannibal, MissouriDowntown Hannibal, Louisiana, MissouriLouisianaEast end of US 61 Bus. overlapMississippi River2.00.0Mark Twain Memorial BridgeIllinoisPike County,_IllinoisPikeLevee Township,_Pike County,_IllinoisLevee Township1 Illinois Route 106IL 106 – Hull, IllinoisHull4 Interstate 172 (Illinois)I-172 north – Quincy, IllinoisQuincyLeft exitKinderhook Township,_Pike County,_IllinoisKinderhook Township10 Illinois Route 96IL 96 / Illinois Route 106IL 106 – Payson, IllinoisPayson, Hull, IllinoisHullBarry,_IllinoisBarry20 To Illinois Route 106IL 106 – Barry, IllinoisBarryNew Salem Township,_Pike County,_IllinoisNew Salem Township31Pittsfield, IllinoisPittsfield, New Salem, IllinoisNew SalemGriggsville Township,_Pike County,_IllinoisGriggsville Township35 U.S. Route 54 in IllinoisUS 54 / Illinois Route 107IL 107 – Pittsfield, IllinoisPittsfield, Griggsville, IllinoisGriggsvilleScott County,_IllinoisScottBloomfield Precinct,_Scott County,_IllinoisBloomfield Precinct46 Illinois Route 100IL 100 – Bluffs, IllinoisBluffs, Detroit, IllinoisDetroitWinchester No. 2 Precinct,_Scott County,_IllinoisWinchester No. 2 Precinct52 Illinois Route 106IL 106 – Winchester, IllinoisWinchesterMorgan County,_IllinoisMorganLynnville Precinct,_Morgan County,_IllinoisLynnville Precinct60 Interstate 72 Business (Jacksonville, Illinois)I-72 Bus. east / U.S. Route 67 in IllinoisUS 67 – Alton, IllinoisAlton, Beardstown, IllinoisBeardstown, Jacksonville, IllinoisJacksonvilleSigned as exits 60A (south) and 60B (north)South Jacksonville,_IllinoisSouth Jacksonville64 Illinois Route 267IL 267 – Alton, IllinoisAlton, Jacksonville, IllinoisJacksonvillePisgah Precinct,_Morgan County,_IllinoisPisgah Precinct68 Interstate 72 Business (Jacksonville, Illinois)I-72 Bus. west to Illinois Route 104IL 104 – Jacksonville, IllinoisJacksonvilleAlexander Precinct,_Morgan County,_IllinoisAlexander Precinct76 Illinois Route 123IL 123 – Ashland, IllinoisAshland, Alexander, IllinoisAlexanderSangamon County,_IllinoisSangamonIsland Grove Township, Sangamon County, IllinoisIsland Grove / New Berlin Township, Sangamon County, IllinoisNew Berlin township line82New Berlin,

IllinoisNew BerlinSpringfield,_IllinoisSpringfield91Wabash Avenue – Springfield, IllinoisSpringfield, Loami, IllinoisLoami93 Illinois Route 4IL 4 – Springfield, IllinoisSpringfield, Chatham, IllinoisChatham96MacArthur Blvd – Springfield, IllinoisSpringfield97 Interstate 55 (Illinois)I-55 south / Interstate 55 Business (Springfield, Illinois)I-55 Bus. north (6th Street) – Springfield, IllinoisSpringfield, St. Louis, MissouriSt. LouisWest end of I-55 overlap; signed as exits 97A (south) and 97B (north)94Stevenson Drive, East Lake Drive96 Illinois Route 29IL 29 (South Grand Avenue) – Taylorville, IllinoisTaylorvilleSigned as exits 96A (south) and 96B (north)98103 Interstate 55 (Illinois)I-55 north – Chicago, IllinoisChicago Illinois Route 97IL 97 west (Clear Lake Avenue)East end of I-55 overlap; signed as exits 98A (east), 98B (west), 103A (south), and 103B (north)Clear Lake Township,_Sangamon County,_IllinoisClear Lake Township104Camp Butler, IllinoisCamp Butler108Riverton, IllinoisRiverton, Dawson, IllinoisDawsonMechanicsburg Township,_Sangamon County,_IllinoisMechanicsburg Township114Buffalo, IllinoisBuffalo, Mechanicsburg, IllinoisMechanicsburg, Dawson, IllinoisDawsonIlliopolis Township,_Sangamon County,_IllinoisIlliopolis Township122Mt. Auburn, IllinoisMt. Auburn, Illiopolis, IllinoisIlliopolisMacon County,_IllinoisMaconNiantic Township,_Macon County,_IllinoisNiantic Township128Niantic, IllinoisNianticHarristown,_IllinoisHarristown133 U.S. Route 36 in IllinoisUS 36 east / U.S. Route 51 in IllinoisUS 51 south – Decatur, IllinoisDecatur, Pana, IllinoisPanaEast end of US 36 overlap; west end of US 51 overlap; signed as exits 133A (east) and 133B (south)Decatur,_IllinoisDecatur138 Illinois Route 121IL 121 – Decatur, IllinoisDecatur, Lincoln, IllinoisLincoln141 U.S. Route 51 in IllinoisUS 51 north / U.S. Route 51 Business (Decatur, Illinois)US 51 Bus. south – Bloomington, IllinoisBloomington, Decatur, IllinoisDecaturEast end of US 51 overlapWhitmore Township,_Macon County,_IllinoisWhitmore Township144 Illinois Route 48IL 48 – Oreana, IllinoisOreana, Decatur, IllinoisDecatur150Argenta, IllinoisArgentaMacon County, IllinoisMacon–Piatt County, IllinoisPiattcounty lineFriends Creek Township, Macon County, IllinoisFriends Creek–Willow Branch Township, Piatt County, IllinoisWillow Branchtownship line156 Illinois Route 48IL 48 – Cisco, IllinoisCisco, Weldon, IllinoisWeldonPiatt County,_IllinoisPiattMonticello Township,_Piatt County,_IllinoisMonticello Township164Monticello, IllinoisMonticelloMonticello Township, Piatt County, IllinoisMonticello–Sangamon Township, Piatt County, IllinoisSangamontownship line166 Illinois Route 105IL 105 west – Monticello, IllinoisMonticelloSangamon Township,_Piatt County,_IllinoisSangamon Township169White Heath Road172 Illinois Route 10IL 10 – Clinton, IllinoisClintonChampaign County,_IllinoisChampaignScott Township,_Champaign County,_IllinoisScott Township176 Illinois Route 47IL 47 – Mahomet, IllinoisMahometChampaign,_IllinoisChampaign182 Interstate 57 (Illinois)I-57 to Interstate 74 (Illinois)I-74 – Memphis, TennesseeMemphis, Chicago, IllinoisChicagoSigned as exits 182A (south) and 182B (north)University Avenue, Church StreetContinuation beyond I-571.000 mi = 1.609 km; 1.000 km = 0.621 mi Concurrency (road)Concurrency terminus • Closed/Former • Incomplete access • UnopenedAuxiliary routesSpur into Quincy, Illinois, Interstate 172I-172Related routesInterstate 72 BusinessLocation:Jacksonville, IllinoisJacksonvilleLength: 9.5 mi Google maps estimate. (15.3 km) Interstate 72 Business (abbreviated BL 72) is a Business Loop of I-72 in Jacksonville, IllinoisJacksonville. It runs from the U.S. Route 36U.S. 36/I-72/U.S. Route 67U.S. 67 interchange southwest of Jacksonville north along the U.S. 67 bypass of Jacksonville to the former alignment of U.S. 36 (Morton Avenue). On Morton Avenue, BL 72 runs east through downtown Jacksonville until it reaches I-72 at exit 68. This is a distance of 9.5 miles (15.3 km).ReferencesExternal links Illinois Highway Ends: Interstate 72

Bement,_Illinois

Bement	
Village	
Country	United States
State	Illinois
County	Piatt
Elevation	686 ft (209.1 m)
Coordinates	39°55′18″N 88°34′23″W
Area	0.8 sq mi (2.07 km^2)
- land	0.8 sq mi (2 km^2)
- water	0 sq mi (0 km^2), 0%
Population	1784 (*2000*)
Density	2197.1 / sq mi (848.3 / km^2)
Timezone	CST (UTC-6)
- summer (DST)	CDT (UTC-5)
ZIP code	61813
Area code	217

Location of Bement within Illinois

Location of Illinois in the United States

Bement is a village in Piatt County, Illinois, in the United States. As of the 2000 census, the village population was 1,784, and in 2009, the population was 1,703.

Geography

Bement is located at 39°55'18"N 88°34'23"W (39.921681, -88.572919)[1].

According to the United States Census Bureau, the village has a total area of 0.8 square miles (2.1 km^2), all of it land.

Demographics

Bement, Illinois Elementary and Middle School.[2]

As of the census[2] of 2000, there were 1,784 people, 687 households, and 485 families residing in the village. The population density was 2,197.1 people per square mile (850.4/km²). There were 723 housing units at an average density of 890.4 per square mile (344.6/km²). The racial makeup of the village was 98.37% White, 0.90% African American, 0.22% Asian, 0.11% from other races, and 0.39% from two or more races. Hispanic or Latino of any race were 0.34% of the population.

There were 687 households out of which 31.7% had children under the age of 18 living with them, 58.5% were married couples living together, 8.7% had a female householder with no husband present, and 29.4% were non-families. 25.6% of all households were made up of individuals and 12.1% had someone living alone who was 65 years of age or older. The average household size was 2.48 and the average family size was 2.97.

In the village the population was spread out with 23.7% under the age of 18, 8.2% from 18 to 24, 29.1% from 25 to 44, 21.8% from 45 to 64, and 17.2% who were 65 years of age or older. The median age was 38 years. For every 100 females there were 96.5 males. For every 100 females age 18 and over, there were 91.8 males.

The median income for a household in the village was $40,163, and the median income for a family was $47,652. Males had a median income of $30,641 versus $21,944 for females. The per capita income for the village was $17,995. About 3.1% of families and 6.2% of the population were below the poverty line, including 5.4% of those under age 18 and 5.7% of those age 65 or over.

Bryant Cottage

Bryant Cottage is located in Bement. Built in 1856 by Francis E. Bryant, it is now preserved as an example of pioneer architecture and as an important historic site. Due to former Governor Blagojevich's budget cuts, Bryant Cottage was scheduled to close, even though it is an important landmark in the political career of one of our nation's greatest Presidents, Abraham Lincoln. But, because of local efforts, Bryant Cottage will remain open to the public. [3]

References

[1] "US Gazetteer files: 2010, 2000, and 1990" (http://www.census.gov/geo/www/gazetteer/gazette.html). United States Census Bureau. 2011-02-12. . Retrieved 2011-04-23.

[2] "American FactFinder" (http://factfinder.census.gov). United States Census Bureau. . Retrieved 2008-01-31.

[3] Maddox, Teri (12/02/2008), "Alton to celebrate 150th anniversary of Lincoln-Douglas debates" (http://www.bnd.com/living/story/ 253793.html) (– Scholar search (http://scholar.google.co.uk/scholar?hl=en&lr=&q=author:Maddox+intitle:Alton+to+celebrate+150th+ anniversary+of+Lincoln-Douglas+debates&as_publication=&as_ylo=&as_yhi=&btnG=Search)), *The Belleville News-Democrat*,

External links

- Bement.com (http://www.bement.com/)
- Bement Centennial 1955 Photographs from the Western Historical Manuscript Collection (http://tjrhino1.umsl. edu/whmc/view.php?description_get=Bement) at the University of Missouri–St. Louis

Illinois_Route_104

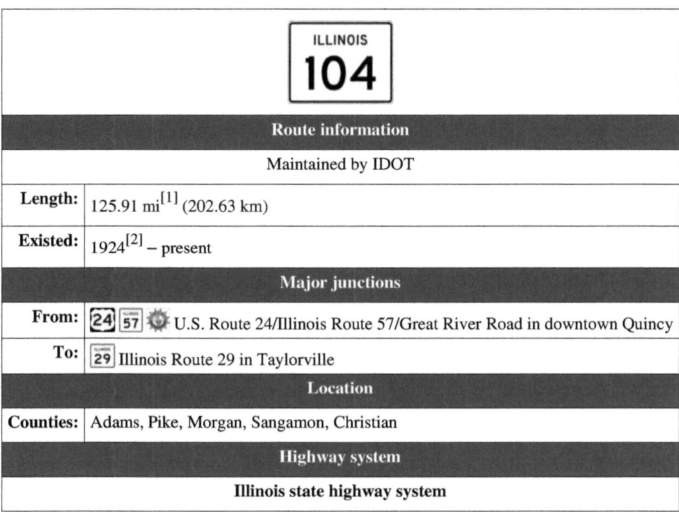

Illinois Route 104 is a state highway in central and western Illinois. It extends from Illinois Route 29 near Taylorville, west over the Illinois River at Meredosia to its western terminus in downtown Quincy. This is a distance of 125.91 miles (202.63 km).[1]

Route description

Illinois 104 crosses Interstate 55 at Exit 82 near Pawnee, and crosses Interstate 172 at Exit 14 near Quincy. The road also crosses Interstate 72 near Jacksonville, but there is no interchange at this crossing.

Illinois 104 doubles as the primary east–west street within the municipality of Quincy, Illinois. Called **Broadway Street**, the highway carries traffic up and down the Mississippi River bluffs that divide the city.

In Jacksonville, Illinois 104 intersects the new U.S. Route 67 Jacksonville Bypass and follows portions of the new Business U.S. 67 through the city. Northwest of the city, Illinois 104 and U.S. 67 are concurrent for 13 miles (21 km).

Points of interest

Points of interest along the road include:

- Illinois College, Jacksonville, Ill., one of the oldest colleges in Illinois (1829) [3].
- Quincy National Cemetery, Quincy, Ill. (1870); 582 interments, including 221 Union soldiers buried before 1882 [4].

History

Prior to 1937, Illinois 104 had run from Mount Sterling on what is now Illinois Route 99 to Taylorville on modern Illinois 104. After 1937, the route took its current routing.[2]

References

[1] Illinois Technology Transfer Center (2006). "T2 GIS Data" (http://www.dot.state.il.us/gist2/select.html). . Retrieved 2007-11-08.

[2] Carlson, Rich. Illinois Highways Page: Routes 101 thru 120 (http://www.n9jig.com/101-120.html). Last updated March 15, 2005.
Retrieved April 28, 2006.

Illinois_Route_47

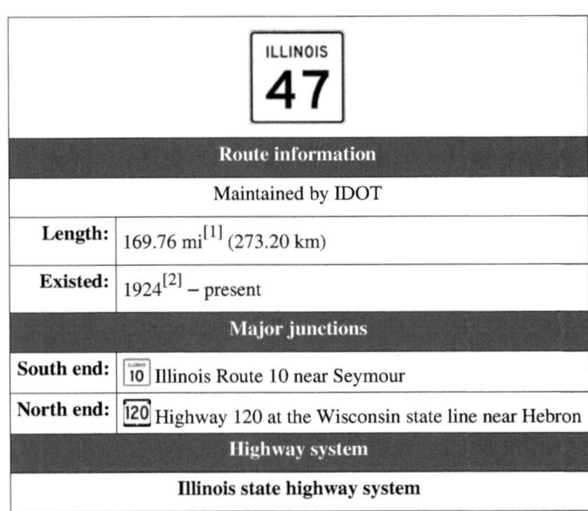

Route information	
Maintained by IDOT	
Length:	169.76 mi[1] (273.20 km)
Existed:	1924[2] – present
Major junctions	
South end:	Illinois Route 10 near Seymour
North end:	Highway 120 at the Wisconsin state line near Hebron
Highway system	
Illinois state highway system	

Illinois Route 47 is a largely rural north–south state highway that runs from the Wisconsin state border at Highway 120 near Hebron, to Illinois Route 10, just south of Interstate 72 near Seymour. This is a distance of 169.76 miles (273.20 km).[1] Even though Route 47 is primarily rural, in several suburbs of Chicago, such as Woodstock, traffic can be congested. Even in heavily rural areas, traffic is heavy.

It crosses most Interstate highways in northern and central Illinois, but the largest towns that Illinois 47 serves are Huntley (at the Jane Addams Memorial Tollway), Yorkville (at U.S. Route 34), Woodstock (at U.S. Route 14), Lily Lake at (Illinois Route 64), Elburn (at Illinois Route 38), Sugar Grove (at U.S. Route 30, Morris (at Interstate 80), Dwight (at Interstate 55), Forrest (at U.S. Route 24), Gibson City (at Illinois Route 54), and Mahomet (at Interstate 74).

Route description

Illinois 47 overlaps Illinois Route 72 and U.S. Route 20 at Pingree Grove, a village approximately 60 miles (97 km) from Chicago; this concurrency is part of a so-called wrong-way concurrency, where one can be driving both *west* on Illinois 72 and *east* on U.S. 20 at the same time. Route 47 also shares concurrencies with Illinois Route 9 and Illinois 54 in Gibson City, U.S. Route 30 in Sugar Grove, U.S. Route 6 in Morris, Historic U.S. Route 66 in Dwight, and U.S. Route 150 in Mahomet.

Future

The Prairie Parkway study is a preliminary study by the Illinois Department of Transportation for a limited-access freeway that will roughly parallel Illinois 47. According to the study's Final Environmental Impact Statement, Illinois Route 47 will interchange with Prairie Parkway approximately 1/2 mile south of Caton Farm Road near Yorkville.

The Illinois Department of Transportation is proposing to add lanes to several portions of Illinois Route 47. A project to reconstruct the road to 4 lanes from Kreutzer Road to Reed Road through Huntley is under construction and scheduled to be completed in October 2011.[3] The Prairie Parkway study included engineering for 4 lanes from I-80 in Morris to Caton Farm Road south of Yorkville. Engineering is underway for Caton Farm Road to Illinois Route 71 in Yorkville, from Kennedy Road in Yorkville to US 30 in Sugar Grove, and from Charles Road to US 14 in Woodstock. A project providing additional lanes through Yorkville is part of the Illinois Jobs Now capital improvement plan and is funded at $36.5 million for construction.[4]

Major junctions

County	Location	Junction	Notes
McHenry		WIS 120	Northern Terminus
	Hebron	IL 173	
	Woodstock	IL 120	
		US 14	
		IL 176	Very short concurrency (0.8 miles)
Kane	Huntley	I-90	No eastbound exit or westbound entrance
	Pingree Grove	US 20 / IL 72	Forms a wrong-way concurrency with US 20 and IL 72 for 0.5 miles
	Lily Lake	IL 64	
	Elburn	IL 38	
		I-88	No eastbound entrance or westbound exit
	Sugar Grove	US 30 / IL 56	North end concurrency with US 30
		US 30	South end of concurrency with US 30
Kendall	Yorkville	US 34	
		IL 126	Western terminus of IL 126
		IL 71	
		US 52	
Grundy	Morris	I-80	Full interchange
	Morris	US 6	Very short concurrency (0.7 miles)
	Mazon	IL 113	Western terminus of IL 113
	Dwight	I-55	Full Interchange

External links

- Illinois Highway Ends: Illinois Route 47 [5]
- IL 47 Improvement Study (Yorkville to Sugar Grove) [6]
- IL 47 Improvement Study (Woodstock) [7]
- IL 47 Corridor Planning Study (Sugar Grove to Hebron) [8]

References

[1] Illinois Technology Transfer Center (2006). "T2 GIS Data" (http://www.dot.state.il.us/gist2/select.html). . Retrieved 2007-11-08.

[2] Carlson, Rick. Illinois Highways Page: Routes 41 thru 60 (http://www.n9jig.com/41-60.html). Last updated March 15, 2006. Retrieved March 28, 2006.

[3] Village of Huntley IL 47 website (http://www.huntley.il.us/Route47Widening.asp) Retrieved 3-9-11

[4] IDOT 2010-2015 Program (http://www.dot.il.gov/hip1015/hwyimprov.html) Retrieved 2-20-10

Macon_County,_Illinois

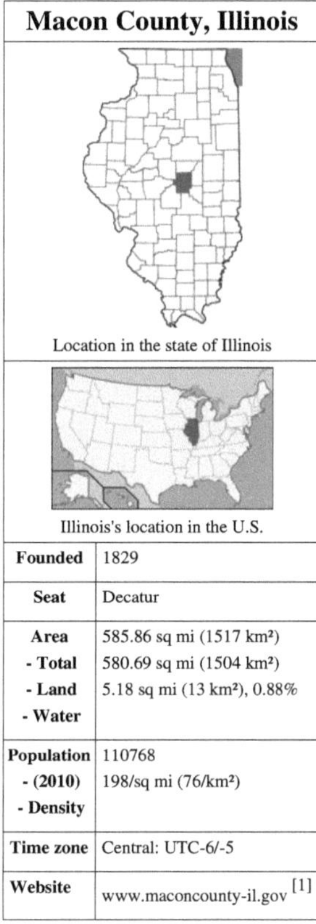

Macon County, Illinois	
Location in the state of Illinois	
Illinois's location in the U.S.	
Founded	1829
Seat	Decatur
Area - **Total** - **Land** - **Water**	585.86 sq mi (1517 km²) 580.69 sq mi (1504 km²) 5.18 sq mi (13 km²), 0.88%
Population - **(2010)** - **Density**	110768 198/sq mi (76/km²)
Time zone	Central: UTC-6/-5
Website	www.maconcounty-il.gov [1]

Macon County is a county located in the U.S. state of Illinois and is included in the Decatur, Illinois, Metropolitan Statistical Area. According to the 2010 census, it has a population of 110,768, which is a decrease of 3.4% from 114,706 in 2000.[1] Its county seat is Decatur.[2]

Geography

According to the 2010 census, the county has a total area of 585.86 square miles (1517.4 km^2), of which 580.69 square miles (1504.0 km^2) (or 99.12%) is land and 5.18 square miles (13.4 km^2) (or 0.88%) is water.[3]

Because of its location, Macon County is often referred to as "The Heart of Illinois" . However, it's not actually in the center of the State.

Major highways

- 72 Interstate 72
- 51 U.S. Highway 51
- 36 U.S. Highway 36
- 48 Illinois Route 48

- [105] Illinois Route 105
- [121] Illinois Route 121
- [128] Illinois Route 128

Townships

Austin, Blue Mound, Decatur, Friends Creek, Harristown, Hickory Point, Illini, Long Creek, Maroa, Milam, Mount Zion, Niantic, Oakley, Pleasant View, South Macon, South Wheatland, Whitmore

Adjacent counties

- De Witt County - north
- Piatt County - northeast
- Moultrie County - southeast
- Shelby County - south
- Christian County - southwest
- Sangamon County - west
- Logan County - northwest

History

Macon County was formed on January 19, 1829 out of Shelby County. It was named for Nathaniel Macon, who served as a Colonel in the Revolutionary War. Macon later served as United States Senator from North Carolina until his resignation in 1828.

Macon County from the time of its creation to 1839

Macon County between 1839 and 1841

Macon County between 1841 and 1843

Macon County in 1843, reduced to its present borders

Demographics

Historical populations		
Census	Pop.	%±
1900	44003	—
1910	54186	23.1%
1920	65175	20.3%
1930	81731	25.4%
1940	84693	3.6%
1950	98853	16.7%
1960	118257	19.6%
1970	125010	5.7%
1980	131375	5.1%
1990	117206	−10.8%
2000	114706	−2.1%
IL Counties 1900-1990 [5]		

As of the census[4] of 2000, there were 114,706 people, 46,561 households, and 30,963 families residing in the county. The population density was 198 people per square mile (76/km²). There were 50,241 housing units at an average density of 86 per square mile (33/km²). The racial makeup of the county was 83.48% White, 14.06% Black or African American, 0.17% Native American, 0.57% Asian, 0.02% Pacific Islander, 0.33% from other races, and 1.36% from two or more races. 0.98% of the population were Hispanic or Latino of any race. 23.8% were of German, 15.9% American, 10.7% English and 9.9% Irish ancestry according to Census 2000.

There were 46,561 households out of which 29.60% had children under the age of 18 living with them, 50.70% were married couples living together, 12.20% had a female householder with no husband present, and 33.50% were non-families. 28.80% of all households were made up of individuals and 11.60% had someone living alone who was 65 years of age or older. The average household size was 2.39 and the average family size was 2.93.

In the county the population was spread out with 24.60% under the age of 18, 9.80% from 18 to 24, 26.40% from 25 to 44, 24.00% from 45 to 64, and 15.20% who were 65 years of age or older. The median age was 38 years. For every 100 females there were 91.20 males. For every 100 females age 18 and over, there were 87.50 males.

The median income for a household in the county was $37,859, and the median income for a family was $47,493. Males had a median income of $39,107 versus $22,737 for females. The per capita income for the county was $20,067. About 9.30% of families and 12.90% of the population were below the poverty line, including 19.00% of those under age 18 and 8.20% of those age 65 or over.

Cities and towns

- Argenta
- Bearsdale
- Blue Mound
- Decatur
- Emery
- Forsyth
- Harristown
- Long Creek
- Macon
- Maroa
- Mount Zion
- Newburg
- Niantic
- Oakley
- Oreana
- Warrensburg

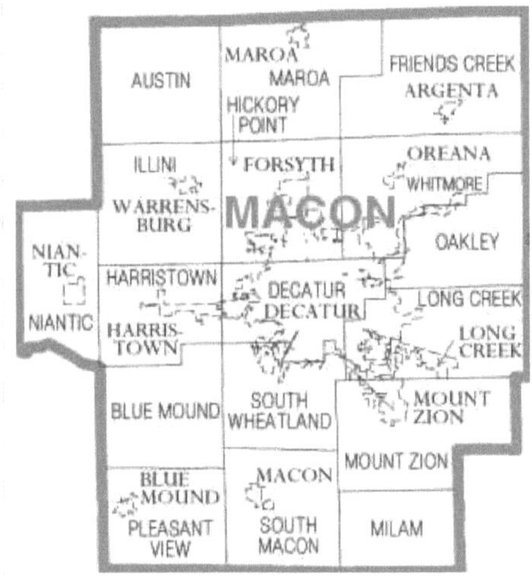

Map of Macon County, Illinois.

Climate and weather

In recent years, average temperatures in the county seat of Decatur have ranged from a low of 17 °F (−8 °C) in January to a high of 88 °F (31 °C) in July, although a record low of −25 °F (−32 °C) was recorded in February 1905 and a record high of 113 °F (45 °C) was recorded in July 1954. Average monthly precipitation ranged from 1.95 inches (50 mm) in February to 4.54 inches (115 mm) in July.[]

See also

- National Register of Historic Places listings in Macon County, Illinois

References

[1] "Macon County QuickFacts" (http://quickfacts.census.gov/qfd/states/17/17115.html). United States Census Bureau. . Retrieved 2011-11-05.
[2] "Find a County" (http://www.naco.org/Counties/Pages/FindACounty.aspx). National Association of Counties. . Retrieved 2011-06-07.
[3] "Census 2010 U.S. Gazetteer Files: Counties" (http://www.census.gov/geo/www/gazetteer/files/Gaz_counties_national.txt). United States Census. . Retrieved 2011-11-05.
[4] "American FactFinder" (http://factfinder.census.gov). United States Census Bureau. . Retrieved 2008-01-31.

Piatt_County,_Illinois

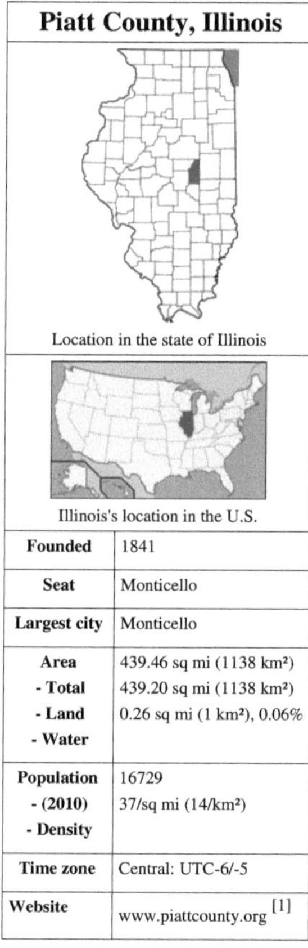

Piatt County, Illinois	
Location in the state of Illinois	
Illinois's location in the U.S.	
Founded	1841
Seat	Monticello
Largest city	Monticello
Area - Total - Land - Water	439.46 sq mi (1138 km²) 439.20 sq mi (1138 km²) 0.26 sq mi (1 km²), 0.06%
Population - (2010) - Density	16729 37/sq mi (14/km²)
Time zone	Central: UTC-6/-5
Website	www.piattcounty.org [1]

Piatt County is a county located in the U.S. state of Illinois. According to the 2010 census, it has a population of 16,729, which is an increase of 2.2% from 16,365 in 2000.[1] Its county seat is Monticello.[2]

Piatt County is part of the Champaign–Urbana Metropolitan Statistical Area.

Geography

According to the 2010 census, the county has a total area of 439.46 square miles (1138.2 km^2), of which 439.20 square miles (1137.5 km^2) (or 99.94%) is land and 0.26 square miles (0.67 km^2) (or 0.06%) is water.[3]

Major highways

- 72 Interstate 72
- 74 Interstate 74
- 36 U.S. Highway 36
- 150 U.S. Highway 150
- 2 Illinois Route 2
- 10 Illinois Route 10
- 32 Illinois Route 32
- 48 Illinois Route 48
- 105 Illinois Route 105

Adjacent counties

- McLean County - north
- Champaign County - east
- Douglas County - southeast
- Moultrie County - south
- Macon County - southwest
- De Witt County - west

History

The first settler was a Quaker named George Haworth, followed by James Martin, Abraham Hanline, Solomon Carter, and William Cordell.

Piatt County was formed in 1841 from Macon and Dewitt Counties. Two local residents: James A. Piatt and Jesse Warner, were instrumental in forming the county. It was named after James A. Piatt after winning a coin flip against Jesse Warner.

Piatt County Illinois Courthouse in Monticello.

Piatt County at the time of its creation in 1841

Abraham Lincoln practiced law in Piatt County as a circuit lawyer. Abraham Lincoln and Stephen A. Douglas planned their presidential debates in Piatt County in 1858. One of which is ornamented by a marker just south of Monticello.

The first courthouse was built in 1843 but replaced by the current courthouse which was finished in 1904.

Illinois Power Company was a major electric utility in Central Illinois, centered in Decatur, Illinois, to the west of Piatt County. At one time, Illinois had a "personal property tax", an ad valorem tax levied by the counties on property that was not real estate. The personal property tax was a major expense for the electric utilities, since their generators and transmission lines were "personal property". Under Illinois law, a corporation, such as Illinois Power, paid personal property tax to the county in which the corporate headquarters was located. Because Piatt County offered a low tax rate, Illinois Power moved its corporate headquarters to that county. This allowed Piatt County to tax utility assets over half of the State, providing a rich source of revenue which was responsible for much of the wealth of this tiny county.

Demographics

Historical populations			
Census	Pop.		%±
1900	17706		—
1910	16376		−7.5%
1920	15714		−4.0%
1930	15588		−0.8%
1940	14659		−6.0%
1950	13970		−4.7%
1960	14960		7.1%
1970	15509		3.7%

1980	16581		6.9%
1990	15548		−6.2%
2000	16365		5.3%
IL Counties 1900-1990 [5]			

As of the census[4] of 2000, there were 16,365 people, 6,475 households, and 4,726 families residing in the county. The population density was 37 people per square mile (14/km²). There were 6,798 housing units at an average density of 15 per square mile (6/km²). The racial makeup of the county was 98.83% White, 0.24% Black or African American, 0.08% Native American, 0.13% Asian, 0.02% Pacific Islander, 0.14% from other races, and 0.57% from two or more races. 0.62% of the population were Hispanic or Latino of any race. 29.9% were of German, 16.1% American, 14.5% English and 13.0% Irish ancestry according to Census 2000.

There were 6,475 households out of which 32.60% had children under the age of 18 living with them, 63.30% were married couples living together, 6.80% had a female householder with no husband present, and 27.00% were non-families. 23.70% of all households were made up of individuals and 11.80% had someone living alone who was 65 years of age or older. The average household size was 2.50 and the average family size was 2.96.

In the county the population was spread out with 25.10% under the age of 18, 6.80% from 18 to 24, 27.60% from 25 to 44, 25.00% from 45 to 64, and 15.50% who were 65 years of age or older. The median age was 40 years. For every 100 females there were 95.40 males. For every 100 females age 18 and over, there were 91.90 males.

The median income for a household in the county was $45,752, and the median income for a family was $52,218. Males had a median income of $36,762 versus $23,606 for females. The per capita income for the county was $21,075. About 3.60% of families and 5.00% of the population were below the poverty line, including 4.80% of those under age 18 and 6.20% of those age 65 or over.

Cities, towns and villages

Cities

- Monticello

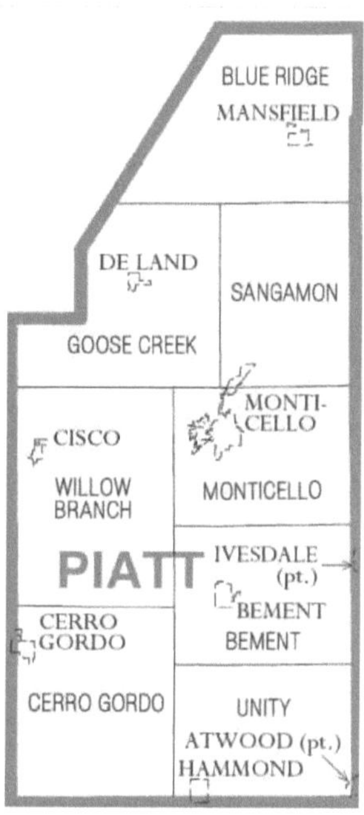

Map of Piatt County, Illinois.

Villages

- Atwood
- Bement
- Cerro Gordo
- Cisco
- De Land
- Hammond
- Mansfield

Unincorporated towns

- Galesville
- La Place
- Lodge
- Milmine
- Pierson Station
- White Heath

Extinct towns

- Blue Ridge
- Centerville
- Combs
- Harris
- Unity
- Voorhies
- Willow Branch

Townships

Piatt County is divided into eight townships:

- Bement
- Blue Ridge
- Cerro Gordo
- Goose Creek
- Monticello
- Sangamon
- Unity
- Willow Branch

Climate and weather

In recent years, average temperatures in the county seat of Monticello have ranged from a low of 14 °F (−10 °C) in January to a high of 85 °F (29 °C) in July, although a record low of −25 °F (−32 °C) was recorded in January 1999 and a record high of 105 °F (41 °C) was recorded in July 1966. Average monthly precipitation ranged from 1.61 inches (41 mm) in January to 3.99 inches (101 mm) in August.[]

See also

- National Register of Historic Places listings in Piatt County

References

[1] "Piatt County QuickFacts" (http://quickfacts.census.gov/qfd/states/17/17147.html). United States Census Bureau. . Retrieved 2011-11-05.
[2] "Find a County" (http://www.naco.org/Counties/Pages/FindACounty.aspx). National Association of Counties. . Retrieved 2011-06-07.
[3] "Census 2010 U.S. Gazetteer Files: Counties" (http://www.census.gov/geo/www/gazetteer/files/Gaz_counties_national.txt). United States Census. . Retrieved 2011-11-05.
[4] "American FactFinder" (http://factfinder.census.gov). United States Census Bureau. . Retrieved 2008-01-31.

Bryant_Cottage_State_Historic_Site

The **Bryant Cottage State Historic Site** is a simple, 1856 four-room house located in Bement, Illinois in the U.S. state of Illinois. It is preserved by the Illinois Historic Preservation Agency as an example of Piatt County, Illinois pioneer architecture and as a key historic site in the 1858 Lincoln-Douglas debates.

Francis E. Bryant

Few settlers moved to southern Piatt County, in east-central Illinois, until the 1850s and the coming of the steam railroad. The ground was flat and open, with few trees to provide firewood for winter. Francis E. Bryant arrived in the young town of Bement in 1856 with a small capital, which he quickly reinvested in general business development as a banker and storekeeper. He bought grain from pioneer farmers, and sold them lumber and coal in return.

Bryant was a member of the U.S. Democratic Party and a personal friend of Senator Stephen A. Douglas, who was running for re-election in 1858.

Stephen A. Douglas

As an incumbent member of the U.S. Senate from Illinois in 1858, Douglas had not expected to have to make a case directly to the people. Under Article I of the U.S. Constitution then in effect, members of the federal Senate were chosen by the state legislatures, not by the voters.

However, Douglas's central position in the growing crisis of American slavery made the election of 1858 extraordinary. Fervent emotions among both Democrats and members of the newly formed Republican party led to a demand that both candidates for the Senate campaign directly among the people of Illinois.

The Republicans nominated Springfield, Illinois lawyer Abraham Lincoln, who wrote to Douglas on July 24, 1858 challenging him to meet and hold a series of nine debates at sites across Illinois. Lincoln renewed this challenge when the two men met in person on the Bement-Monticello road, now Illinois Route 105, on July 29.

The campaigning Douglas was at the time going southward to Bement, where he would spend the night in the Bryant Cottage. It was during this one-night stay that Douglas decided to accept most of Lincoln's challenge and debate him seven times. Douglas wrote Lincoln from Bement on the morning of July 30, suggesting that the two men debate in Ottawa, Illinois, Freeport, Illinois, Jonesboro, Illinois, Charleston, Illinois, Galesburg, Illinois, Quincy, Illinois, and Alton, Illinois.

Lincoln accepted these terms in a letter dated July 31.

Abraham Lincoln

At the time of the Lincoln-Douglas challenge of July 1858, Stephen A. Douglas was an experienced, incumbent U.S. Senator; Abraham Lincoln was a lawyer in private practice with little successful office-holding experience. Illinois Republicans had nominated him to face Douglas because of his skill at making speeches and his ability to frame the issue of slavery in a manner that conveyed visceral opposition to the institution without antagonizing racist American voters.

Douglas had every reason to avoid Lincoln's challenge to debate. As the incumbent, he had an advantage in terms of name recognition. The senator chose to voluntarily debate his challenger because he agreed with Lincoln that slavery was a growing crisis and because he had faith that his own doctrine of popular sovereignty would create a solution to the dilemma.

At first, Douglas appeared to be the winner in this challenge. His legislative candidates defeated the Republicans in November 1858, thereby assuring Douglas's re-election as U.S. senator. However, Lincoln and Douglas re-visited the issue of slavery in the U.S. presidential election of 1860. And in this final contest between the two men, Lincoln was the victor.

Local folklore

After the deaths of both Douglas and Lincoln, the Bryant family came to believe that the two men had met in person in the parlor of the Bryant Cottage to negotiate the details of their debates. It is clear that Douglas made the key decision, that of accepting Lincoln's challenge, in the Bryant Cottage, but the surviving letters between the two men appear to indicate that they negotiated on paper after their face-to-face roadside meeting of July 29, 1858.

Current status

In response to state budget cuts, the state of Illinois temporarily closed the Bryant Cottage State Historic Site from October 2008 until April 2009. The site reopened to the public on April 23, 2009.

External links

- State of Illinois Bryant Cottage site [1]
- Bryant Cottage State Historic Site [2]
- Lincoln and Douglas at the Bryant Cottage [3]

Illinois

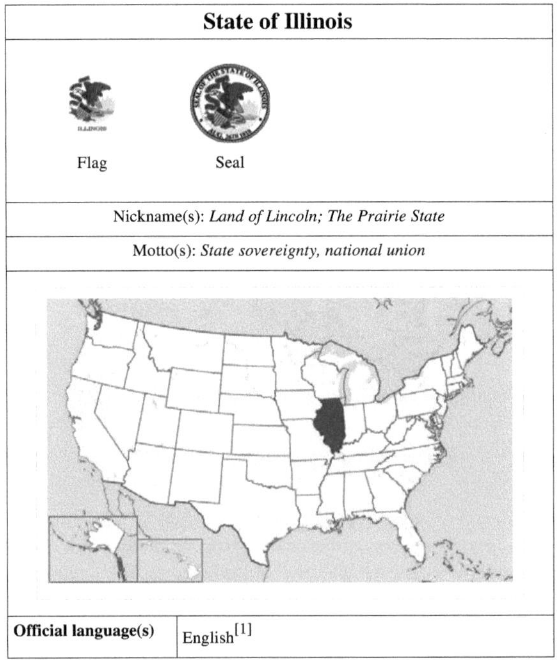

State of Illinois	
Flag	Seal
Nickname(s): *Land of Lincoln; The Prairie State*	
Motto(s): *State sovereignty, national union*	
Official language(s)	English[1]

Spoken language(s)	English (80.8%) Spanish (10.9%) Polish (1.6%) Other (6.7%)[2]
Demonym	Illinoisan
Capital	Springfield
Largest city	Chicago
Largest metro area	Chicago metropolitan area
Area	Ranked 25th in the U.S.
- Total	57,914 sq mi (149,998 km^2)
- Width	210 miles (340 km)
- Length	395 miles (629 km)
- % water	4.0/ Negligible
- Latitude	36° 58′ N to 42° 30′ N
- Longitude	87° 30′ W to 91° 31′ W
Population	Ranked 5th in the U.S.
- Total	12,830,632 (2010)[3]
- Density	223.4/sq mi (86.27/km^2) Ranked 12th in the U.S.
- Median income	$54,124[4] (17)
Elevation	
- Highest point	Charles Mound[5] [6] [7] 1,235 ft (376.4 m)
- Mean	600 ft (180 m)
- Lowest point	Confluence of Mississippi River and Ohio River[6] [7] 280 ft (85 m)
Admission to Union	December 3, 1818 (21st)
Governor	Pat Quinn (D)
Lieutenant Governor	Sheila Simon (D)
Legislature	General Assembly
- Upper house	Senate
- Lower house	House of Representatives
U.S. Senators	Dick Durbin (D) Mark Kirk (R)
U.S. House delegation	11 Republicans, 8 Democrats (list)
Time zone	Central: UTC-6/-5
Abbreviations	IL, Ill., US-IL
Website	[www.illinois.gov www.illinois.gov]

Illinois (◀)) ᶦ/ˌɪlɪˈnɔɪ/ *il-i-noy*) is the fifth-most populous state of the United States of America, and is often noted for being a microcosm of the entire country.[8] With Chicago in the northeast, small industrial cities and great agricultural productivity in central and northern Illinois, and natural resources like coal, timber, and petroleum in the south, Illinois has a broad economic base. Illinois is a major transportation hub. The Port of Chicago connects the state to other global ports from the Great Lakes, via the St. Lawrence Seaway, to the Atlantic Ocean; as well as the Great Lakes to the Mississippi River, via the Illinois River. For decades, O'Hare International Airport has ranked as one of the world's busiest airports. As the "most average state",[8] Illinois has long had a reputation as a bellwether both in social and cultural terms[8] and politics.

In the 1810s, settlers began arriving from Kentucky. In 1818 Illinois achieved statehood. The state's population originally grew from south to north. Chicago was founded in the 1830s on the banks of the Chicago River, one of the few natural harbors on southern Lake Michigan.[9] Railroads and John Deere's invention of the self-scouring steel plow turned Illinois' rich prairie into some of the world's most productive and valuable farmlands, attracting immigrant farmers from Germany and Sweden. By 1900, the growth of industrial jobs in the northern cities and coal mining in the central and southern areas attracted immigrants from Eastern and Southern Europe. Illinois was an important manufacturing center during both world wars. The Great Migration established a large community of African Americans in Chicago that created the city's famous jazz and blues cultures.[10] [11]

Three U.S. Presidents have been elected while living in Illinois—Abraham Lincoln, Ulysses S. Grant, and Barack Obama. Additionally, President Ronald Reagan, whose political career was based in California, was the only US President actually born and raised in Illinois. Today, Illinois honors Lincoln with its official state slogan, *Land of Lincoln*, which has been displayed on its license plates since 1954.[12] [13]

Name

"Illinois" is the modern spelling for the early French missionaries and explorers' name for the Illinois people, a name that was spelled in many different ways in the early records.[14]

The name "Illinois" has traditionally been said to mean "man" or "men" in the Miami-Illinois language, with the original *iliniwek* transformed via French into Illinois.[15] [16] However, this etymology is not supported by the Illinois language itself, in which the word for 'man' is *ireniwa* and plural 'men' is *ireniwaki*. The name *Illiniwek* has also been said to mean "tribe of superior men",[17] though this is nothing more than a false etymology. In fact the name "Illinois" derives from the Miami-Illinois verb *irenwe·wa* "he speaks the regular way". This was then taken into the Ojibwe language, perhaps in the Ottawa dialect, and modified into *ilinwe·* (pluralized as *ilinwe·k*). These forms were then borrowed into French, where the /we/ ending acquired the spelling *-ois*. The current form, *Illinois*, began to appear in the early 1670s. The Illinois's name for themselves, as attested in all three of the French missionary-period dictionaries of Illinois, was *Inoka*, of unknown meaning and unrelated to the other terms.[18] [19] [20]

History

Pre-European

Native Americans lived along the waterways of the Illinois area for thousands of years before the arrival of Europeans. The Koster Site has been excavated and demonstrated 7,000 years of continuous habitation. Cahokia, the largest regional chiefdom and urban center of the Pre-Columbian Mississippian culture, was located near present-day Collinsville, Illinois. They built more than 100 mounds and a woodhenge in a planned design expressing the culture's cosmology. The civilization vanished in the 15th century for unknown reasons, but historians and archeologists have speculated that the people depleted the area of resources. Many indigenous tribes engaged in constant warfare. According to Suzanne Austin Alchon, "At one site in the central Illinois River valley, one-third of all adults died as a result of violent injuries."[21]

Copper plates found at Pre-Columbian burial sites in Illinois.

The next major power in the region was the Illinois Confederation or Illini, a political alliance among several tribes. There were about 25,000 Illinois Indians in 1700, but systematic attacks and warfare by the Iroquois reduced their numbers by 90%.[22] Gradually, members of the Potawatomi, Miami, Sauk, and other tribes came in from the east and north.[23] In the American Revolution, the Illinois and Potawatomi supported the American colonists' cause.

European exploration

Illinois in 1718, approximate modern state area highlighted, from *Carte de la Louisiane et du cours du Mississipi* by Guillaume de L'Isle.[24]

French explorers Jacques Marquette and Louis Jolliet explored the Illinois River in 1673. In 1680, other French explorers constructed a fort at the site of present day Peoria, and in 1682, a fort atop Starved Rock in today's Starved Rock State Park. As a result of this French exploration, Illinois was part of the French empire until 1763, when it passed to the British with their conquest of New France. The small French settlements continued; a few British soldiers were posted in Illinois, but there were no British or American settlers. In 1778, George Rogers Clark claimed the Illinois Country for Virginia. The area was ceded by Virginia to the new United States in 1783 and became part of the Northwest Territory.[25]

19th century

Historical populations		
Census	**Pop.**	**%±**
1800	2458	—
1810	12282	399.7%
1820	55211	349.5%
1830	157445	185.2%
1840	476183	202.4%
1850	851470	78.8%
1860	1711951	101.1%
1870	2539891	48.4%
1880	3077871	21.2%
1890	3826352	24.3%
1900	4821550	26.0%
1910	5638591	16.9%
1920	6485280	15.0%
1930	7630654	17.7%
1940	7897241	3.5%
1950	8712176	10.3%
1960	10081158	15.7%
1970	11113976	10.2%
1980	11426518	2.8%
1990	11430602	0%
2000	12419293	8.6%
2010	12830632	3.3%
Source: 1910–2010[26]		

The Illinois-Wabash Company was an early claimant to much of Illinois. The Illinois Territory was created on February 3, 1809, with its capital at Kaskaskia.

During the discussions leading up to Illinois' admission to the Union, the proposed northern boundary of the state was moved twice.[27] The original provisions of the Northwest Ordinance had specified a boundary that would have been tangent to the southern tip of Lake Michigan. Such a boundary would have actually left Illinois with no shoreline on Lake Michigan at all. However, as Indiana had successfully been granted a 10-mile northern extension of its boundary to provide it with a usable lakefront, the original bill for Illinois statehood, submitted to Congress on January 23, 1818, stipulated a northern border at the same latitude as Indiana's which is defined as 10 miles (16 km) north of the southernmost extremity of Lake Michigan. But the Illinois delegate, Nathaniel Pope, wanted more. Pope lobbied to have the boundary moved further north, and the final bill passed by Congress did just such; it included an amendment to shift the border to 42° 30' north, which is approximately 51 miles (82 km) north of the Indiana northern border. This shift added 8500 square miles (22000 km^2) to the state, including the lead mining region near Galena. More importantly, it added nearly 50 miles of Lake Michigan shoreline and the Chicago River. Pope and others envisioned a canal which would connect the Chicago and Illinois rivers, and thusly, connect the Great Lakes to the Mississippi.

In 1818, Illinois became the 21st U.S. state. The capital remained at Kaskaskia, headquartered in a small building rented by the state. In 1819, Vandalia became the capital, and over the next 18 years, three separate buildings were built to serve successively as the capitol building. In 1837, the state legislators representing Sangamon County, under the leadership of state representative Abraham Lincoln, succeeded in having the capital moved to Springfield,[28] where a fifth capitol building was constructed. A sixth capitol building was erected in 1867, which continues to serve as the Illinois capitol today.

Though ostensibly a "free state", Illinois had slavery. The French owned black slaves as late as the 1820s. Slavery was nominally banned by the Northwest Ordnance, but that was not enforced. When Illinois became a sovereign state in 1818, the Ordnance no longer applied, and there were about 900 slaves there. As the southern part of the state, known as "Egypt", was largely settled by migrants from the South, the section was hostile to free blacks and allowed settlers to bring slaves with them for labor. Most citizens were opposed to allowing blacks as permanent residents, and efforts to make slavery official failed in 1822. Nevertheless, some slaves were brought in seasonally or as house servants.[29] The Illinois Constitution of 1848 was written with a provision for exclusionary laws to be passed. In 1853, John A. Logan helped pass a law to prohibit all African Americans, including freedmen, from settling in the state.[30]

In 1832, the Black Hawk War was fought in Illinois and current day Wisconsin between the United States and the Sauk, Fox (Meskwaki) and Kickapoo Indian tribes. The Indians withdrew to Iowa; when they attempted to return, they were defeated by U.S. militia and forced back to Iowa.

The winter of 1830–1831 is called the "Winter of the Deep Snow"; a sudden, deep snowfall blanketed the state, making travel impossible for the rest of the winter, and many travelers perished. Several severe winters followed, including the "Winter of the Sudden Freeze". On December 20, 1836, a fast-moving cold front passed through, freezing puddles in minutes and killing many travelers who could not reach shelter. The adverse weather resulted in crop failures in the northern part of the state. The southern part of the state shipped food north and this may have contributed to its name: "Little Egypt", after the Biblical story of Joseph in Egypt supplying grain to his brothers.[31]

By 1839, the Mormons had founded a utopian city called Nauvoo. Located in Hancock County, along the Mississippi River, Nauvoo flourished and soon rivaled Chicago for the position of the state's largest city. But in 1844, the Mormon leader Joseph Smith was murdered in the Carthage Jail, about 30 miles away from Nauvoo. Soon afterward, after close to six years of rapid development, Nauvoo saw a rapid decline after the Mormons' new leadership led them out of Illinois in a mass exodus to present-day Utah.

Chicago gained prominence as a Great Lakes port and then as an Illinois and Michigan Canal port after 1848, and as a rail hub soon afterward. By 1857, Chicago was Illinois' largest city.[25] With the tremendous growth of mines and factories in the state in the 19th century, Illinois played an important role in the formation of labor unions in the United States. The Pullman Strike and Haymarket Riot in particular greatly influenced the development of the American labor movement. From Sunday, October 8, 1871, until Tuesday, October 10, 1871, the Great Chicago Fire burned in downtown Chicago, destroying 4 square miles (10 km^2).[32]

In 1847, after lobbying by Dorothea L. Dix, Illinois became one of the first states to establish a system of state-supported treatment of mental illness and disabilities, replacing local almshouses.

Civil War

. During the American Civil War, over 250,000 Illinois men served in the Union Army, a figure surpassed by only New York, Pennsylvania, and Ohio. Beginning with President Abraham Lincoln's first call for troops and continuing throughout the war, Illinois mustered 150 infantry regiments, which were numbered from the 7th to the 156th regiments. Seventeen cavalry regiments were also gathered, as well as two light artillery regiments.[33] The town of Cairo at the southern tip of the state served as a strategically important supply base and training center for the Union army. For several months, both General Grant and Admiral Foote had headquarters in Cairo.

Embarkation of Union troops from Cairo on January 10, 1862

20th century

At the turn of the 20th century, Illinois had a population of nearly 5 million. Bolstered by continued immigration from southern and eastern Europe, and by African-Americans from Mississippi, Louisiana, and Arkansas, Illinois grew and emerged as one of the most important states in the union. By the end of the century, the population had reached 12.4 million.

The Century of Progress World's Fair was held at Chicago in 1933. Oil strikes in Marion County and Crawford County lead to a boom in 1937, and, by 1939, Illinois ranked fourth in U.S. oil production. Chicago became an ocean port with the opening of the Saint Lawrence Seaway in 1959. The seaway and the Illinois Waterway connected Chicago to both the Mississippi River and the Atlantic Ocean. In 1960, Ray Kroc opened the first McDonald's franchise in Des Plaines (which still exists today as a museum, with a working McDonald's across the street).

No state has had a more prominent role than Illinois in the emergence of the nuclear age. As part of the Manhattan Project, the first sustained nuclear chain reaction took place at the University of Chicago in 1942. In 1957, Argonne National Laboratory, near Chicago, activated the first experimental nuclear power generating system in the United States. By 1960, the first privately financed nuclear plant in United States, Dresden 1, was dedicated near Morris. In 1967, Fermilab, a national nuclear research facility near Batavia, opened a particle accelerator which was the world's largest for over 40 years. And, with eleven plants currently operating, Illinois leads all states in the amount of electricity generated from nuclear power.[34]

The state's fourth constitution was adopted in 1970, replacing the 1870 document. The first Farm Aid concert was held in Champaign to benefit American farmers, in 1985. The worst upper Mississippi River flood of the century, the Great Flood of 1993, inundated many towns and thousands of acres of farmland.[25]

Geography

Boundaries

Illinois' eastern border with Indiana consists of a north-south line at 87° 31' 30" west longitude, from Lake Michigan to the Wabash River, above Post Vincennes. The Wabash River continues as the eastern/southeastern border with Indiana, until the Wabash enters the Ohio River. This marks the beginning of Illinois' southern border with Kentucky, which runs along the northern shoreline of the Ohio River.[35] Its western border with Missouri and Iowa is the Mississippi River. Its northern border with Wisconsin is fixed at 42° 30' north latitude. The northeastern border of

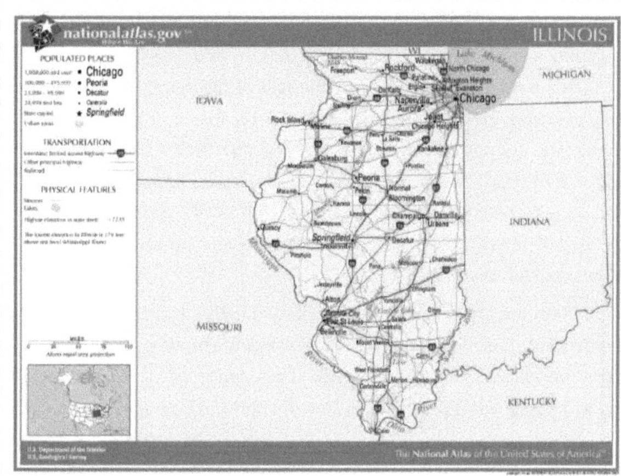

Illinois, showing major cities and roads

Illinois actually lies within Lake Michigan, within which Illinois shares a water boundary with the state of Michigan.[23]

Topography

Though Illinois lies entirely in the Interior Plains, it does have some minor variation in its elevation. In extreme northwestern Illinois, the Driftless Area, a region of unglaciated and therefore higher and more rugged topography, occupies a small part of the state. Charles Mound, located in this region, has the state's highest elevation above sea level at 1,235 feet (376 m). The floodplain on the Mississippi River from Alton to the Kaskaskia River is known as the American Bottom.

Divisions

Illinois has three major geographical divisions. Northern Illinois is dominated by the Chicago metropolitan area; the city of Chicago, its suburbs, and the adjoining exurban area into which the metropolis is expanding. As defined by the federal government, the Chicago metro area includes several counties in Illinois, Indiana, and Wisconsin. Chicago is a cosmopolitan city, densely populated, industrialized, the transportation hub of the nation, and settled by a wide variety of ethnic groups with a population of 9.8 million people. The city of Rockford, the fourth largest metropolitan area, and the state's third largest city, sits along Interstates 39 and 90 some 75 miles (121 km) northwest of Chicago. The Quad Cities region, located along the Mississippi River in northern Illinois, had a population of 379,066 in 2009.

Southward and westward, the second major division is Central Illinois, an area of mostly prairie. Known as the Heart of Illinois, it is characterized by small towns and mid-sized cities. The western section (west of the Illinois River) was originally part of the Military Tract of 1812 and forms the conspicuous western bulge of the state. Agriculture, particularly corn and soybeans, as well as educational institutions and manufacturing centers, figure prominently. Cities include Peoria, the third largest metropolitan area in Illinois at 370,000; Springfield, the state capital; Quincy; Decatur; Bloomington-Normal; and Champaign-Urbana.[23]

The third division is Southern Illinois, comprising the area south of U.S. Route 50, including Little Egypt, near the juncture of the Mississippi River and Ohio River. Southern Illinois is the site of the ancient city of Cahokia, as well as the site of the first state capital at Kaskaskia, which today is separated from the rest of the state by the Mississippi River.[23] [36] This region can be distinguished from the other two by its warmer climate, different variety of crops (including some cotton farming in the past), more rugged topography (due to the area remaining unglaciated during the Illinoian Stage, unlike most of the rest of the state), as well as small-scale oil deposits and coal mining. The Illinois suburbs of St. Louis comprise the second most populous metropolitan area in Illinois with over 700,000 inhabitants, and are known collectively as the Metro-East. The other significant concentration of population in Southern Illinois is the Carbondale-Marion-Herrin, Illinois Combined Statistical Area centered on Carbondale and Marion, a two-county area that is home to 123,272 residents.[23] A portion of southeastern Illinois is part of the extended Evansville, Indiana Metro Area, locally referred to as the Tri-State with Indiana and Kentucky. Seven Illinois counties are in the area.

In addition to these three, largely latitudinally defined divisions, all of the region outside of the Chicago Metropolitan area is often called "downstate" Illinois. This term is flexible, but is generally meant to mean everything outside the Chicago-area. Thus, some cities in *Northern* Illinois, such as DeKalb, which is west of Chicago, and Rockford—which is actually *north* of Chicago—are considered to be "downstate".

Climate

Because of its nearly 400 mile distance from its northernmost and southernmost extremes, as well as its mid-continental situation, Illinois has a widely varying climate. Most of Illinois has a humid continental climate (Köppen climate classification *Dfa*), with hot, humid summers and cold winters. The southernmost part of the state, from about Carbondale southward, borders on a humid subtropical climate (Koppen *Cfa*), with more moderate winters. Average yearly precipitation for Illinois varies from just over 48 inches (1219 mm) at the southern tip to around 35 inches (889 mm) in the northern portion of the state. Normal annual snowfall exceeds 38 inches (965 mm) in the Chicago area, while the southern portion of the state normally receives less than 14 inches (356 mm).[37] The all time high temperature was 117 °F (47 °C), recorded on July 14, 1954, at East St. Louis, Illinois, while the all time low temperature was −36 °F (−38 °C), recorded on January 5, 1999, at Congerville, Illinois.[38]

Illinois averages around 51 days of thunderstorm activity a year, which ranks somewhat above average in the number of thunderstorm days for the United States. Illinois is vulnerable to tornadoes with an average of 35 occurring annually, which puts much of the state at around five tornadoes per 10000 square miles (30000 km^2) annually.[39] The deadliest tornadoes on record in the nation have occurred largely in Illinois, not because the tornadoes are more common or frequent in Illinois, but rather, because Illinois is simply the most populous state in Tornado Alley. The Tri-State Tornado of 1925 killed 695 people in three states; 613 of the victims died in Illinois.[40] Modern developments in storm tracking have caused death tolls from tornadoes to dramatically decline since the 1960s, with no major losses of life in the state since the 1967 tornado storm in northern Illinois.

City	Jan	Feb	Mar	Apr	May	Jun	Jul	Aug	Sep	Oct	Nov	Dec
Cairo[41]	41/25	47/29	57/39	69/50	77/58	86/67	90/71	88/69	81/61	71/49	57/39	46/30
Chicago[42]	30/16	36/21	47/30	59/40	71/51	81/61	85/65	83/65	75/57	64/45	48/34	36/22
Edwardsville[43]	36/19	42/24	52/34	64/45	75/55	84/64	89/69	86/66	79/58	68/46	53/35	41/25
Moline[44]	30/12	36/18	48/29	62/39	73/50	83/60	86/64	84/62	76/53	64/42	48/30	34/18
Peoria[45]	31/14	37/20	49/30	62/40	73/51	82/60	86/65	84/63	77/54	64/42	49/31	36/20
Rockford[46]	27/11	33/16	46/27	59/37	71/48	80/58	83/63	81/61	74/52	62/40	46/29	32/17
Springfield[47]	33/17	39/22	51/32	63/42	74/53	83/62	86/66	84/64	78/55	67/44	51/34	38/23

Demographics

In the 2010 Census, Illinois was found to have a population of 12,830,632, representing a slightly lower increase than had been anticipated in estimates from two years earlier. Illinois is the most populous state in the Midwest region. Chicago, the third most populous city in the United States, is the center of the Chicago metropolitan area. *Chicagoland*, as this area is known locally, comprises only 8% of the land area of the state, but contains 65% of the state's residents.

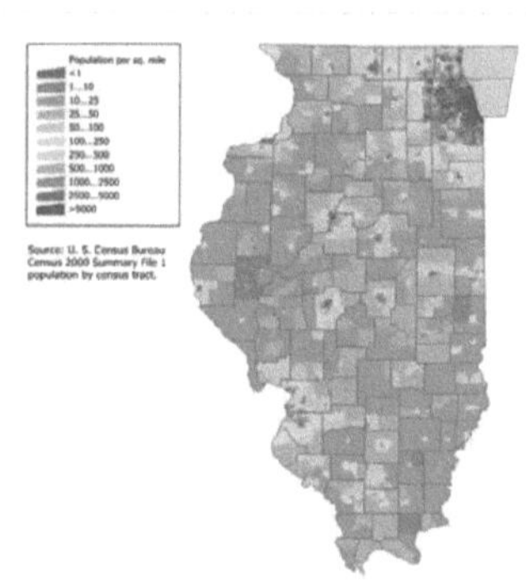

Illinois Population Density Map

Specific demographic data from the 2010 Census is not subject to release until March 2011, but as of the 2007 estimates from the U.S. Census Bureau, there were 1,768,518 foreign-born inhabitants of the state or 13.8% of the population, with 48.4% from Latin America, 24.6% from Asia, 22.8% from Europe, 2.9% from Africa, 1.2% from Northern America and 0.2% from Oceania. Of the foreign-born population, 43.7% were naturalized U.S. citizens and 56.3% were not U.S. citizens.[48] Additionally, the racial distributions were as follows: 65.0% White American, 15.0% African American, 14.9% Latino American, 4.3% Asian American, 0.3% American Indian and Alaska Natives, and 0.1% Native Hawaiians and Pacific Islander American.[49] In 2007, 6.9% of Illinois' population was reported as being under age 5, 24.9% under age 18 and 12.1% were age 65 and over. Females made up approximately 50.7% of the population.[49]

According to the 2007 estimates, 21.1% of the population had German ancestry, 13.3% had Irish ancestry, 7.9% had Polish ancestry, 6.7% had English ancestry, 6.4% had Italian ancestry, 4.6% listed themselves as American, 2.4%

had Swedish ancestry, 2.2% had French ancestry, other than Basque, 1.6% had Dutch ancestry, 1.4% had Norwegian ancestry and 1.3% had Scottish ancestry.[48] Also, 21.8% of the population age 5 years and over reported speaking a language other than English, with 12.8% of the population speaking Spanish, 5.6% speaking other Indo-European languages, 2.5% speaking Asian and Austronesian languages, and 0.8% speaking other languages.[48]

Chicago, along the shores of Lake Michigan, is the nation's third largest city. In 2000, 23.3% of Illinois' population lived in the city of Chicago, 43.3% in Cook County and 65.6% in the counties of the Chicago metropolitan area: Will, DuPage, Kane, Lake and McHenry counties, as well as Cook County. The remaining population lives in the smaller cities and rural areas that dot the state's plains. As of 2000, the state's center of population was at 41°16′42″N 88°22′49″W, located in Grundy County, northeast of the village of Mazon.[23] [25] [36] [50]

Largest cities				
Rank	City	Population[51]	County	
1	Chicago	2,695,598	Cook, DuPage	Chicago
2	Aurora	197,899	DuPage Kane Kendall Will	
3	Rockford	152,871	Winnebago	
4	Joliet	147,433	Will	Springfield (State Capital)
5	Naperville	141,853	DuPage, Will	
6	Springfield	117,352	Sangamon	
7	Peoria	115,007	Peoria	
8	Elgin	108,188	Cook, Kane	
9	Waukegan	89,078	Lake	
10	Cicero	85,616	Cook	
11	Champaign	81,055	Champaign	
12	Arlington Heights	76,031	Cook	
based on 2010 U.S. Census				

Urban areas

Chicago is the largest city in the state and the third most populous city in the United States, with its 2010 population of 2,695,598. The U.S. Census Bureau currently lists seven other cities with populations of over 100,000 within Illinois. Based upon the Census Bureau's official 2010 population,[52] : Aurora, a Chicago satellite town which eclipsed Rockford for the title of "Second City" of Illinois in 2006; its 2010 population was 197,899. Rockford, at 152,871, is the third largest city in the state, and is also the largest city in the state not located within the Chicago metropolitan area. Joliet, located southwest of Chicago, is the fourth largest city in the state, with a population of 147,433. Naperville, a suburb of Chicago, is fifth with 141,853; Naperville and Aurora (the 2nd largest city) share a boundary along Illinois Route 59. Springfield, the state capital of Illinois, comes in sixth with 117,352. Peoria, which decades ago was the second largest city in the state, comes in seventh with 115,007. The eighth largest and final city in the 100,000 club is Elgin, a northwest suburb of Chicago with a 2010 population of 108,188.

The most populated city in the state south of Springfield is Belleville, with 44,478 people at the 2010 census. It is located in the Illinois portion of Greater St. Louis (often called the Metro-East area), which has a rapidly growing population of over 700,000 people.

Other major urban areas include the Champaign-Urbana Metropolitan Area, which has a combined population of almost 230,000 people, the Illinois portion of the Quad Cities area with about 215,000 people, and the Bloomington-Normal area with a combined population of over 165,000.

Religion

Roman Catholics constitute the single largest religious denomination in Illinois; they are heavily concentrated in and around Chicago, and account for nearly 30% of the state's population.[53] However, taken together *as a group*, the various Protestant denominations comprise a greater percentage of the state's population than do Catholics. In 2000 Catholics in Illinois numbered 3,874,933, the largest Protestant denominations were the United Methodist Church, with 365,182 members, and the Southern Baptist Convention, with 305,838. Jews constituted the largest non-Christian group with 270,000 adherents.[54] Chicago and its suburbs are also home to a large and growing population of Hindus, Muslims, Baha'is and Sikhs.

Illinois played an important role in the early Latter Day Saint movement, with Nauvoo, Illinois, becoming a gathering place for Mormons in the early 1840s. Nauvoo was the location of the succession crisis, which led to the separation of the Mormon movement into several Latter Day Saint sects. The Church of Jesus Christ of Latter-day Saints, the largest of the sects to emerge from the Mormon schism, claims 55,460 in Illinois today.[55]

Demographics of Illinois (csv) [56]

By race	White	Black	AIAN*	Asian	NHPI*
2000 (total population)	80.71%	15.73%	0.62%	3.84%	0.11%
2000 (Hispanic only)	11.78%	0.35%	0.19%	0.08%	0.04%
2005 (total population)	80.34%	15.63%	0.62%	4.45%	0.11%
2005 (Hispanic only)	13.72%	0.39%	0.20%	0.09%	0.04%
Growth 2000–05 (total population)	2.30%	2.07%	3.74%	19.16%	10.13%
Growth 2000–05 (non-Hispanic only)	-0.68%	1.81%	0.91%	19.36%	10.18%
Growth 2000–05 (Hispanic only)	19.75%	13.28%	10.14%	9.96%	10.06%
* AIAN is American Indian or Alaskan Native; NHPI is Native Hawaiian or Pacific Islander					

Economy

The dollar gross state product for Illinois was estimated to be US$652 billion in 2010.[56] The state's 2010 per capita gross state product was estimated to be US$45,302,[56] and the state's per capita personal income was estimated to be US$41,411 in 2009.[57]

As of March 2010, the state's unemployment rate was 11.5%,[58] which fell to 9.9% by August 2011.[59]

Taxes

Illinois' state income tax is calculated by multiplying net income by a flat rate. In 1990, that rate was set at 3%, but in 2010, the General Assembly voted in a temporary increase in the rate to 5%; the new rate

The Federal Reserve Bank of Chicago at the heart of Chicago's financial center

went into effect on January 1, 2011, and is scheduled to return to 3% after four years.[60] [61] There are two rates for state sales tax: 6.25% for general merchandise and 1% for qualifying food, drugs and medical appliances.[62] The property tax is the largest single tax in Illinois, and is the major source of tax revenue for local government taxing districts. The property tax is a local—not state—tax, imposed by local government taxing districts, which include counties, townships, municipalities, school districts and special taxation districts. The property tax in Illinois is imposed only on real property.[23] [25] [36]

Agriculture

Illinois' major agricultural outputs are corn, soybeans, hogs, cattle, dairy products, and wheat. In most years, Illinois is either the first or second state for the highest production of soybeans, with a harvest of 427.7 million bushels (11.64 million metric tons) in 2008, after Iowa's production of 444.82 million bushels (12.11 million metric tons).[63] Illinois ranks second in U.S. corn production with more than 1.5 billion bushels produced annually.[64] Illinois also produces wine, and the state is home to two American viticultural areas. Illinois' universities are actively researching alternative agricultural products as alternative crops.

Manufacturing

Illinois is one of the nation's manufacturing leaders, boasting annual value added productivity by manufacturing of over $107 billion in 2006. About three-quarters of the state's manufacturers are located in the Northeastern Opportunity Return Region, with 38 percent of Illinois' approximately 18,900 manufacturing plants located in Cook County. As of 2006, the leading manufacturing industries in Illinois, based upon value-added, were chemical manufacturing ($18.3 billion), machinery manufacturing ($13.4 billion), food manufacturing ($12.9 billion), fabricated metal products ($11.5 billion), transportation equipment ($7.4 billion), plastics and rubber products ($7.0 billion), and computer and electronic products ($6.1 billion).[65]

Services

By the early 2000s, Illinois' economy had moved toward a dependence on high-value-added services, such as financial trading, higher education, law, logistics, and medicine. In some cases, these services clustered around institutions that hearkened back to Illinois' earlier economies. For example, the Chicago Mercantile Exchange, a trading exchange for global derivatives, had begun its life as an agricultural futures market. Other important non-manufacturing industries include publishing, tourism, and energy production and distribution.

Energy

Illinois is a net importer of fuels for energy, despite large coal resources and some minor oil production. Illinois exports electricity, ranking fifth among states in electricity production and seventh in electricity consumption.[66]

Coal

The coal industry of Illinois has its origins in the middle 19th century, when entrepreneurs such as Jacob Loose discovered coal in locations such as Sangamon County. Jacob Bunn contributed to the development of the Illinois coal industry, and was a founder and owner of the Western Coal & Mining Company of Illinois. About 68% of Illinois has coal-bearing strata of the Pennsylvanian geologic period. According to the Illinois State Geological Survey, 211 billion tons of bituminous coal are estimated to lie under the surface, having a total heating value greater than the estimated oil deposits in the Arabian Peninsula.[67] However, this coal has a high sulfur content, which causes acid rain unless special equipment is used to reduce sulfur dioxide emissions.[23] [25] [36] Many Illinois power plants are not equipped to burn high-sulfur coal. In 1999, Illinois produced 40.4 million tons of coal, but only 17 million tons (42%) of Illinois coal was consumed in Illinois. Most of the coal produced in Illinois is exported to other states, while much of the coal burned for power in Illinois (21 million tons in 1998) is mined in the Powder River Basin of Wyoming.[66]

Mattoon was recently chosen as the site for the Department of Energy's FutureGen project, a 275 megawatt experimental zero emission coal-burning power plant which just received a second round of funding from the DOE.

Petroleum

Illinois is a leading refiner of petroleum in the American Midwest, with a combined crude oil distillation capacity of nearly 900000 barrels per day (m^3/d). However, Illinois has very limited crude oil proved reserves that account for less than 1% of U.S. crude oil proved reserves. Residential heating is 81% natural gas compared to less than 1% heating oil. Illinois is ranked 14th in oil production among states, with a daily output of approximately 28000 barrels (4500 m^3) in 2005.[68]

Nuclear power

Nuclear power arguably began in Illinois with the Chicago Pile-1, the world's first artificial self-sustaining nuclear chain reaction in the world's first nuclear reactor, built on the University of Chicago campus. There are six operating nuclear power plants in Illinois: Braidwood; Byron; Clinton; Dresden; LaSalle; and Quad Cities. With the exception of the single-unit Clinton plant, each of these facilities has two reactors. Three reactors have been permanently shut down and are in various stages of decommissioning: Dresden-1 and Zion-1 and 2. As of 2008, Illinois was ranked first among the 50 states both in nuclear capacity and nuclear generation.[69] In 2007, 48% of Illinois' electricity was generated using nuclear power.[69]

Byron Nuclear Generating Station in Ogle County.

Wind power

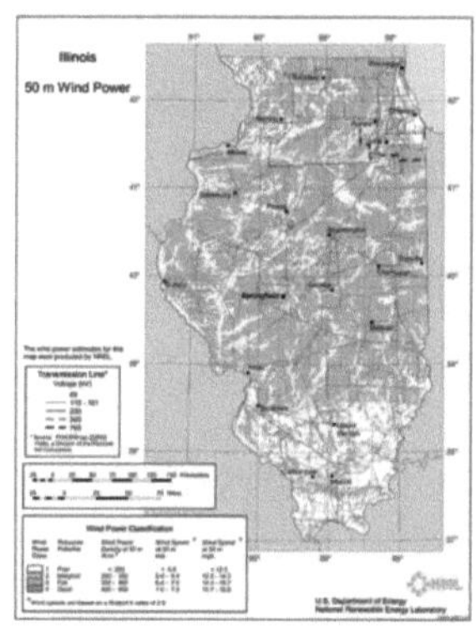

Average annual wind power distribution for Illinois, 50 m (160 ft) height above ground (2009).

Illinois has seen growing interest in the use of wind power for electrical generation.[70] Most of Illinois was rated in 2009 as "marginal or fair" for wind energy production by the U.S. Department of Energy, with some western sections rated "good" and parts of the south rated "poor".[71] These ratings are for wind turbines with 50-metre (160 ft) hub heights; newer wind turbines are taller, enabling them to reach stronger winds farther from the ground. As a result, more areas of Illinois have become prospective wind farm sites. As of September 2009, Illinois had 1116.06 MW of installed wind power nameplate capacity with another 741.9 MW under construction.[72] Illinois ranked ninth among U.S. states in installed wind power capacity, and sixteenth by potential capacity.[72] Large wind farms in Illinois include Twin Groves, Rail Splitter, EcoGrove, and Mendota Hills.[72]

As of 2007, wind energy represented only 1.7% of Illinois' energy production, and it was estimated that wind power could provide 5–10% of the state's energy needs.[73] [74] Also, the Illinois General Assembly mandated in 2007 that by 2025, 25% of all electricity generated in Illinois is to come from renewable resources.[75]

Biofuels

Illinois is ranked second in corn production among U.S. states, and Illinois corn is used to produce 40% of the ethanol consumed in the United States.[64] The Archer Daniels Midland corporation in Decatur, Illinois is the world's leading producer of ethanol from corn.

University of Illinois at Urbana-Champaign is one of the partners in the Energy Biosciences Institute (EBI), a $500 million biofuels research project funded by petroleum giant BP.[76] [77]

Arts and culture

Museums

For a more comprehensive list, see List of museums in Illinois.

Illinois has numerous museums; the greatest concentration of these is in Chicago. Numerous museums in the city of Chicago are considered some of the best in the world. These include the John G. Shedd Aquarium, the Field Museum of Natural History, the Art Institute of Chicago, the Adler Planetarium, and the Museum of Science and Industry.

Magnolia Manor is a Victorian period historic house museum in Cairo.

The state of the art Abraham Lincoln Presidential Library and Museum in Springfield is the largest presidential library in the country. Other historical museums in the state include Magnolia Manor in Cairo, the Elihu Benjamin Washburne and Ulysses S. Grant Homes, both in Galena, and the Polish Museum of America in Chicago.

Music

Illinois is a leader in music education having hosted the Midwest Clinic: An International Band and Orchestra Conference since 1946, as well being home to the Illinois Music Educators Association (IMEA), one of the largest professional music educator's organizations in the country. Each summer since 2004, Southern Illinois University Carbondale has played host to the Southern Illinois Music Festival, which presents dozens of performances throughout the region. Past featured artists include the Eroica Trio and violinist David Kim.

Sports

For a more comprehensive list, see List of professional sports teams in Illinois.

Major league teams

As one of the United States' major metropolises, all major sports leagues have teams headquartered in Chicago.

Soldier Field, Chicago

- Two Major League Baseball teams are located in the state. The Chicago Cubs of the National League play in the second-oldest major league stadium (Wrigley Field) and are widely known for having the longest championship drought in all of major American sport: not winning the World Series since 1908.[78] [79] The Chicago White Sox of the American League won the World Series in 2005, their first since 1917.

- The Chicago Bears football team has won nine total NFL Championships, the last occurring in Super Bowl XX in 1986.
- The Chicago Bulls of the NBA is one of the most recognized basketball teams in the world, due largely to the efforts of Michael Jordan, who led the team to six NBA championships in eight seasons in the 1990s.
- The Chicago Blackhawks of the NHL began playing in 1926, as a member of the Original Six and have won four Stanley Cups, most recently in 2010.
- The Chicago Fire soccer club is a member of MLS and is one of the league's most successful and best-supported, since its founding in 1997, winning one league and four Lamar Hunt U.S. Open Cups in that timespan.
- The Chicago Carnage of the MLRH is the most recent professional team in Chicago.

Minor league teams

Many minor league teams also call Chicago their home. These include

- The Chicago Rush of the Arena Football League, who won ArenaBowl XX in 2006.
- The Chicago Wolves is an AHL team.
- The Chicago Sky of the WNBA
- The Chicago Bandits of the NPF, a female softball league; the Bandits won their first title in 2008
- The Chicago Red Stars of the WPS, a female soccer league

Former Chicago sports franchises

Folded teams

The city was formerly home to several other teams that either failed to survive, or that belonged to leagues that folded.

- The Chicago Blitz, United States Football League
- The Chicago Sting, Major Indoor Soccer League
- The Chicago Cougars, World Hockey Association
- The Chicago Rockers, Continental Basketball Association
- The Chicago Skyliners, American Basketball Association
- The Chicago Bruisers, Arena Football League
- The Chicago Power, National Professional Soccer League
- The Chicago Blaze, National Women's Basketball League.

Relocated teams

The NFL's Arizona Cardinals, who currently play in Phoenix, Arizona, played in Chicago as the Chicago Cardinals, until moving to St. Louis, Missouri after the 1959 season.

Professional sports teams outside of Chicago

Chicago is not the only place in Illinois where professional sports are played. The Rockford Lightning is one of the oldest CBA teams in the league. The Peoria Chiefs and Kane County Cougars are minor league baseball teams affiliated with MLB. The Schaumburg Flyers and Lake County Fielders are members of the North American League, and the Southern Illinois Miners, Gateway Grizzlies, Joliet Slammers, Windy City ThunderBolts and Normal CornBelters belong to the Frontier League.

In addition to the Chicago Wolves, the AHL also has two teams in Illinois outside of Chicago: the Rockford IceHogs serves as the AHL affiliate of the Chicago Blackhawks, and the Peoria Rivermen is the AHL affiliate of the St. Louis Blues.

Motor racing

Illinois has a long tradition of motor racing. Oval tracks at the Chicagoland Speedway in Joliet, the Chicago Motor Speedway in Cicero and the Gateway International Raceway in Madison, near St. Louis, have hosted NASCAR, CART and IRL races, whereas the Sports Car Club of America, among other national and regional road racing clubs, have visited the Autobahn Country Club in Joliet, the Blackhawk Farms Raceway in South Beloit and the former Meadowdale International Raceway in Carpentersville. Illinois also has several short tracks and dragstrips. The dragstrip at Gateway International Raceway and the Route 66 Raceway, which sits on the same property as the Chicagoland Speedway, both host NHRA drag races.

Parks and recreation

For a more comprehensive list, see List of protected areas of Illinois.

The Illinois state parks' system began in 1908 with what is now Fort Massac State Park, becoming the first park in a system encompassing over 60 parks and about the same number of recreational and wildlife areas.

Areas under the protection and control of the National Park Service include: the Illinois and Michigan Canal National Heritage Corridor near Lockport;[80] the Lewis and Clark National Historic Trail; the Lincoln Home National Historic Site in Springfield; the Mormon Pioneer National Historic Trail; the Trail of Tears National Historic Trail; and the American Discovery Trail.[81]

In March 2011, Illinois ranked as a bottom-seven "Worst" state (tied with Georgia and Oklahoma) in the American State Litter Scorecard. The Land of Lincoln suffers from overall poor effectiveness and quality of its statewide public space cleanliness—due to state and related eradication standards and performance indicators.[82]

Governance

While the organization of the central government of Illinois is largely the same as every other state (having three branches of government: executive, legislative and judicial), below this top level, the substructure of Illinois' government is extremely complex, arguably the most complex of all fifty states.

State government structure

Legislative functions are granted to the Illinois General Assembly, composed of the 118-member Illinois House of Representatives and the 59-member Illinois Senate. The executive branch is led by the Governor of Illinois, but four other executive officials are separately elected by the people. The judiciary is composed of the Supreme Court of Illinois and the lower appellate and circuit courts.[35]

Illinois' uniquely complex local government structure

Illinois has more units of local government than any other state—over 8,000 in all.[83] The basic subdivision of Illinois is like almost every other state, the county, and Illinois has 102 of these. About half of these counties, in turn, are divided into townships, which is much the same as many other Midwestern states. And finally, Illinois has a number of cities, villages, and towns commensurate with a state of its size. But the counties, townships, and municipal governments in Illinois make up only about 1/4 of all of the governmental units in the state. The reason Illinois has so many units of government is because so many single-purpose governmental entities have been created. The following is just a partial list of the types of single-purpose governmental units in Illinois.

- Illinois has school districts which do not share boundaries with either counties nor townships. While this is not unique to Illinois, what *would* strike observers from many other states as odd, is that *there are many places where a given piece of land sits within **two** school districts*—one high school district, and another elementary district—each of which has its own school board and its own taxing authority.

- Another common political unit is the **library district**.[84] Library districts are run by library boards; such boards are elected bodies and have the power to levy taxes in their district.[85] Library boards in some districts are elected at the general election, but in other districts may be held in conjunction with local elections, or even, as stand-alone elections.[86] The boundaries of these library districts occasionally coincide with those of another governmental entity, such as a township, but more often, they are set independently.
- Another unit of government with taxation authority is the **sanitary district**,[87] a euphemism for "sewage district". (Many Illinoisians first learned of the existence of these entities when, in 1978, a sanitary district board member named Alex Seith captured the Democratic nomination for U.S. Senate against the veteran senator Charles Percy and nearly upset him in the general election.) The largest of the sanitary districts in the state is the Metropolitan Water Reclamation District of Greater Chicago (*né Sanitary District of Chicago*), which oversaw the reversal of the course of the Chicago River.

There are additional units of government that oversee watersheds, land use, and many other functions that in another state would be handled by the county or city governments.

Homerule

The Constitution of 1970 created, for the first time in Illinois, a type of "home rule", which allows cities of certain sizes to opt out of certain types of state laws.

Law enforcement

For a more comprehensive list, see List of law enforcement agencies in Illinois.

The complexity and overlapping jurisdictions of Illinois' law enforcement agencies is not unlike that of the overlapping taxing authorities noted above. At the state level, there are at least eleven law enforcement agencies. At the county level, there are sheriffs, forest preserve police and other specialized police forces. At the local level, most cities and many villages have municipal police forces, park district police forces, and even local specialized police forces. Many colleges also have their own campus police that are often sworn police officers.

In 2000, Illinois was ranked 4th in the U.S. in the number of full-time sworn officers with 321 per 100,000 persons, behind Louisiana (415), New York (384), and New Jersey (345).[88] In this ranking, only New York had a higher total population than Illinois. Illinois is also near the top of most law enforcement numbers lists, such as number of agencies per state, number of agencies with special jurisdictions, and number of local police agencies.[88] Even taking into account that Illinois is the fifth most populous state, many of the ratios are higher than more populated states. There is much overlap in jurisdiction amongst the different law enforcement agencies.

Politics

Party balance

Historically, Illinois was long a major swing state, with near-parity existing between the Republican and the Democratic parties. However, in recent elections, the Democratic Party has slowly gained ground, and Illinois has come to be seen as more of a "blue" state.[89] [90] Chicago and most of Cook County votes have long been strongly Democratic. In addition, Democratic voters have moved to the traditionally Republican "collar counties" (the suburbs surrounding Chicago's Cook County, Illinois), which are becoming increasingly diverse.[91] [92]

The dome on the Illinois State Capitol in Springfield is taller than the dome on the United States Capitol.

Republicans continue to prevail in rural northern and central Illinois; Republican support is strong in southern Illinois outside of the East St. Louis metropolitan area. Illinois has voted for Democratic presidential candidates in the last five elections. State resident Barack Obama easily won the state's 21 electoral votes in 2008, by a margin of 25 percentage points with 61.9% of the vote. And though the Republicans' electoral performance in the 2010 mid-year elections marked some improvement, the trend in Illinois politics for the long term appears to be more blue than red.[93] [94]

History of corruption

Politics in the state, particularly those of the Chicago machine, have been famous for highly visible corruption cases, as well as for crusading reformers, such as governors Adlai Stevenson (D) and James R. Thompson (R). In 2006, former Governor George Ryan (R) was convicted of racketeering and bribery. In 2008, then-Governor Rod Blagojevich (D) was served with a criminal complaint on corruption charges, stemming from allegations that he conspired to sell the vacated Senate seat left by President Barack Obama (D) to the highest bidder. In the late 20th century, Congressman Dan Rostenkowski (D) was imprisoned for mail fraud; former governor and federal judge Otto Kerner, Jr. (D) was imprisoned for bribery; and State Auditor of Public Accounts (Comptroller) Orville Hodge (R) was imprisoned for embezzlement. In 1912, William Lorimer, the GOP boss of Chicago, was expelled from the U.S. Senate for bribery and in 1921, Governor Len Small (R) was found to have defrauded the state of a million dollars.[25] [36] [95]

US Presidents from Illinois

Three presidents have claimed Illinois as their political base: Lincoln, Grant, and Obama. Lincoln was born in Kentucky, but moved to Illinois at the age of 21; he served in the General Assembly and represented the 7th congressional district in the US House of Representatives before his election as President. Ulysses S. Grant was born in Ohio and had a military career that precluded settling down, but on the eve of the Civil War, and approaching middle age, Grant moved to Illinois and thus claimed it as his home when running for President. Barack Obama was born and raised in Hawaii (other than a four year period of his childhood spent in Indonesia) and made Illinois his home and base after completing law school.

Only one person elected President of the United States was actually born in Illinois. Ronald Reagan was born in Tampico, raised in Dixon and educated at Eureka College. Reagan moved to Los Angeles as a young adult and later became Governor of California before being elected President.

Though never elected president, Illinois Governor Adlai Stevenson, who was born and raised in central Illinois, was the Democratic nominee for president in 1952 and 1956.

Black senators

Since the adoption of the United States Constitution in 1789, only six African-Americans have served as members of the United States Senate, and *half* of them represented Illinois: Carol Moseley-Braun, Barack Obama,[96] and Roland Burris, who was appointed to replace Obama after his election to the presidency.

Education

Illinois State Board of Education

The Illinois State Board of Education (**ISBE**) is autonomous of the governor and the state legislature, and administers public education in the state. Local municipalities and their respective school districts operate individual public schools but the ISBE audits performance of public schools with the Illinois School Report Card. The ISBE also makes recommendations to state leaders concerning education spending and policies.

Primary and secondary schools

Education is compulsory from ages 7 to 17 in Illinois. Schools are commonly but not exclusively divided into three tiers of primary and secondary education: elementary school, middle school or junior high school and high school. District territories are often complex in structure. Many areas in the state are actually located in *two* school districts—one for high school, the other for elementary and middle schools. And such districts do not necessarily share boundaries. A given high school may have several elementary districts that feed into it, yet some of those feeder districts may themselves feed into multiple high school districts.

Colleges and universities

For a more comprehensive list, see List of colleges and universities in Illinois.

Using the criterion established by the Carnegie Foundation for the Advancement of Teaching, there are eleven "National Universities" in the state. As of 19 August 2010, five of these rank in the "first tier" (that is, the top quartile) among the top 500 National Universities in the United States, as determined by the *U.S. News & World Report* rankings: the University of Chicago (8), Northwestern University (12), the University of Illinois at Urbana-Champaign (39), Illinois Institute of Technology (106), and Loyola University Chicago (119).[97]

Illinois also has more than 20 additional accredited four-year universities, both public and private, and dozens of small liberal arts colleges across the state. Additionally, Illinois supports 49 public community colleges in the Illinois Community College System.

Infrastructure

Transportation

Because of its central location and its proximity to the Rust Belt and Grain Belt, Illinois is a national crossroads for air, auto, rail and truck traffic.

Airports

From 1962 until 1998, Chicago's O'Hare International Airport (ORD) was the busiest airport in the world, measured both in terms of total flights and passengers. While it was surpassed by Atlanta's Hartsfield in 1998, with 59.3 million domestic passengers annually, along with

1980s Illinois license plate, displaying the **Land of Lincoln** slogan that has been featured on the state's plates since 1954.

11.4 million international passengers in 2008,[98] O'Hare remains one of the two or three busiest airports in the world, and some years still ranks number one in total flights. It is a major hub for United Airlines and American Airlines, and a major airport expansion project is currently underway. Chicago Midway International Airport (MDW), which had been the busiest airport in the world until supplanted by O'Hare in 1962, is now the secondary airport in the Chicago metropolitan area. For a time in the late 1960s and 1970s, Midway was nearly vacant except for general aviation, but growth in the area, combined with political deadlock over the building of a new major airport in the region, has caused a resurgence for Midway. It is now a major hub for Southwest Airlines, and services many other airlines as well. Midway served 17.3 million domestic and international passengers in 2008.[99]

Rail

Illinois has an extensive passenger and freight rail transportation network. Chicago is a national Amtrak hub and in-state passengers are served by Amtrak's Illinois Service, featuring the Chicago to Carbondale *Illini* and *Saluki*, the Chicago to Quincy *Carl Sandburg* and *Illinois Zephyr*, and the Chicago to St. Louis *Lincoln Service*. Currently there is trackwork on the Chicago-St. Louis line to bring the maximum speed up to 110 mph (180 km/h) which would reduce the trip time by an hour and a half. Nearly every North American railway meets at Chicago, making it the largest and most active rail hub in the country. Extensive commuter rail is provided in the city proper and some immediate suburbs by the Chicago Transit Authority's 'L' system. The largest suburban commuter rail system in the United States, operated by Metra, uses existing rail lines to provide direct commuter rail access for hundreds of suburbs to the city and beyond.

In addition to the state's rail lines, the Mississippi River and Illinois River provide major transportation routes for the state's agricultural interests. Lake Michigan gives Illinois access to the Atlantic Ocean by way of the Saint Lawrence Seaway.

Interstate highway system

Illinois' central location and large population are the reasons that Illinois carries the distinction of having the most primary (2-digit) Interstates pass through it among the 50 states.

Major U.S. Interstate highways crossing the state include: I-24, I-39, I-55, I-57, I-64, I-70, I-72, I-74, I-80, I-88, I-90, and I-94.

See also

- List of National Register of Historic Places in Illinois
- List of people from Illinois

References

[1] "(5 ILCS 460/20) (from Ch. 1, par. 2901-20) State Designations Act." (http://www.ilga.gov/legislation/ilcs/ilcs3.asp?ActID=132&
ChapAct=5 ILCS 460/&ChapterID=2&ChapterName=GENERAL+PROVISIONS&ActName=State+Designations+Act.).
Illinois Compiled Statutes. Springfield, Illinois: Illinois General Assembly. September 4, 1991. . Retrieved April 10, 2009. "Sec. 20. Official
language. The official language of the State of Illinois is English."

[2] "Illinois Table: QT-P16; Language Spoken at Home: 2000" (http://factfinder.census.gov/servlet/QTTable?_bm=y&
-geo_id=04000US17&-qr_name=DEC_2000_SF3_U_QTP16&-ds_name=DEC_2000_SF3_U&-_lang=en&-redoLog=false&
-CONTEXT=qt). *Data Set: Census 2000 Summary File 3 (SF 3) – Sample Data*. U.S. Census Bureau. 2000. . Retrieved April 10, 2009.

[3] "Annual Estimates of the Resident Population for the United States, Regions, States, and Puerto Rico: April 1, 2000 to July 1, 2009" (http://
www.census.gov/popest/states/tables/NST-EST2009-01.csv). United States Census Bureau. . Retrieved January 4, 2010.

[4] "Median Household Income (In 2007 Inflation-Adjusted Dollars) Universe: Households" (http://factfinder.census.gov/servlet/
GRTTable?_bm=y&-geo_id=01000US&-_box_head_nbr=R1901&-ds_name=ACS_2007_1YR_G00_&-_lang=en&-redoLog=false&
-format=US-30&-mt_name=ACS_2007_1YR_G00_R1902_US30&-CONTEXT=grt). *2007 American Community Survey 1-Year Estimates*.
U.S. Census Bureau. 2007. . Retrieved April 9, 2009.

[5] "Charles" (http://www.ngs.noaa.gov/cgi-bin/ds_mark.prl?PidBox=NJ0855). *NGS data sheet*. U.S. National Geodetic Survey. . Retrieved
October 20, 2011.

[6] "Elevations and Distances in the United States" (http://egsc.usgs.gov/isb/pubs/booklets/elvadist/elvadist.html). United States
Geological Survey. 2001. . Retrieved October 21, 2011.

[7] Elevation adjusted to North American Vertical Datum of 1988.

[8] Ohlemacher, Stephen (May 17, 2007). "Analysis ranks Illinois most average state" (http://www.southernillinoisan.com/articles/2007/05/
17/top/20300809.txt). Associated Press. Carbondale, Illinois: The Southern Illinoisan. . Retrieved April 10, 2009.

[9] "Chicago's Front Door: Chicago Harbor." A digital exhibit published online by the Chicago Public Library. (http://www.chipublib.org/
digital/lake/CFDHarbor.html). Retrieved October 20, 2007.

[10] Jazz (http://encyclopedia.chicagohistory.org/pages/665.html). Encyclopedia of Chicago. Retrieved on October 5, 2011.

[11] Blues (http://encyclopedia.chicagohistory.org/pages/151.html). Encyclopedia of Chicago. Retrieved on October 5, 2011.

[12] http://www.cyberdriveillinois.com/special/plate_history/start_history.html

[13] "Slogan" (http://www.museum.state.il.us/exhibits/symbols/slogan.html). Museum.state.il.us. . Retrieved February 7, 2011.

[14] Fay, J. (2009) *Eriniouaj*. Retrieved October 21, 2009 from http://www.illinoisprairie.info/Eriniouaj.htm.

[15] Hodge, Frederick Webb (1911). *Handbook of American Indians north of Mexico, Volume 1* (http://books.google.com/
?id=ze4YAAAAYAAJ&pg=PA597). Smithsonian Institution, Bureau of American Ethnology. p. 597. OCLC 26478613. .

[16] Stewart, George R. (1967) [1945]. *Names on the Land: A Historical Account of Place-Naming in the United States* (Sentry (3rd) ed.).
Houghton Mifflin.

[17] "Illinois Symbols" (http://www.illinois.gov/facts/symbols.cfm). State of Illinois. . Retrieved April 20, 2006.

[18] Callary, Edward (2008). *Place Names of Illinois* (http://books.google.com/?id=ZvHgwa-XImcC&pg=PA169). University of Illinois
Press. p. 169. ISBN 9780252033568. .

[19] Costa, David J. (January 2007). "Three American Placenames: Illinois" (http://myaamia.strackattack.com/OtherFiles/CostaNewsletter.
pdf#page=9). *Society for the Study of the Indigenous Languages of the Americas Newsletter* **25** (4): 9–12. ISSN 1046-4476. . Retrieved May
29, 2011.

[20] Fay, J. (2010) *"*Inoka" or "Inoka:" an Internet Troll or Gag*. Retrieved December 23, 2010 from http://www.illinoisprairie.info/Inoka.
htm.

[21] Austin Alchon, Suzanne (2003). *A pest in the land: new world epidemics in a global perspective* (http://books.google.com/
books?id=YiHHnV08ebkC&pg=PA59&dq#v=onepage&q=&f=false). University of New Mexico Press. p. 59. ISBN 0826328717. .

[22] Frederick E. Hoxie, *Encyclopedia of North American Indians* (1996) 266-7, 506

[23] Nelson, Ronald E. (ed.), ed (1978). *Illinois: Land and Life in the Prairie State*. Dubuque, Iowa: Kendall/Hunt. ISBN 0-8403-1831-6.

[24] de L'Isle, Guillaume (1718). "*Carte de la Louisiane et du Cours du Mississipi. 1718.*" (http://www2.lib.virginia.edu/exhibits/
lewis_clark/exploring/ch2-10.html). *An Exhibition of Maps and Navigational Instruments on View*. Tracy W. McGregor Room, Alderman
Library: University of Virginia. . Retrieved January 25, 2010.

[25] Biles, Roger (2005). *Illinois: A History of the Land and its People*. DeKalb: Northern Illinois University Press. ISBN 0-87580-349-0.

[26] Resident Population Data. "Resident Population Data – 2010 Census" (http://2010.census.gov/2010census/data/apportionment-pop-text.
php). 2010.census.gov. . Retrieved February 7, 2011.

[27] "Full Remarks from Dave M" (http://www.sancohis.org/presentations/Illinois From Territory to State.htm). Sancohis.org. March 16,
2010. . Retrieved February 7, 2011.

[28] "Abraham Lincoln and Springfield – Abraham Lincoln's Classroom" (http://abrahamlincolnsclassroom.org/Library/newsletter.
asp?ID=137&CRLI=193). Abrahamlincolnsclassroom.org. . Retrieved February 7, 2011.

[29] Paul Finkelman, *Slavery and the founders: race and liberty in the age of Jefferson* (2001) p 78

[30] James Pickett Jones, *Black Jack: John A. Logan and Southern Illinois in the Civil War Era* 1967 ISBN 0-8093-2002-9.

[31] Duff, Judge Andrew D. Egypt (http://www.springhousemagazine.com/egypt2.htm). Republished, *Springhouse Magazine*, accessed May
1, 2006.

[32] Roland Tweet, *Miss Gale's Books: The Beginnings of the Rock Island Public Library*, (Rock Island, IL: Rock Island Public Library, 1997), 15.

[33] Illinois in the Civil War. Illinois Infantry, Cavalry, and Artillery Units (http://www.illinoiscivilwar.org/units_num.html). Retrieved November 26, 2006.

[34] "ComEd and Electricity Related Messages for Economic Development" (https://www.comed.com/sites/PartnersBusiness/Documents/EconomicDevelopmentFactSheet.pdf) (PDF). . Retrieved February 7, 2011.

[35] Wikisource. Illinois Constitution of 1818.

[36] Horsley, A. Doyne (1986). *Illinois: A Geography*. Boulder: Westview Press. ISBN 0-86531-522-1.

[37] Illinois State Climatologist Office (http://www.sws.uiuc.edu/atmos/statecli/index.htm). Climate Maps for Illinois (http://www.sws.uiuc.edu/atmos/statecli/Mapsv2/mapsv2.htm). Retrieved April 22, 2006.

[38] NWS Chicago, IL (November 2, 2005). "Public Information Statement" (http://www.crh.noaa.gov/product.php?site=LOT&product=PNS&issuedby=LOT). . Retrieved January 15, 2010.

[39] " Annual average number of tornadoes, 1953–2004 (http://www.ncdc.noaa.gov/img/climate/research/tornado/small/avgt5304.gif)", NOAA National Climatic Data Center. Retrieved on October 24, 2006.

[40] PAH Webmaster (November 2, 2005). "NWS Paducah, KY: NOAA/NWS 1925 Tri-State Tornado Web Site – General Information" (http://www.crh.noaa.gov/pah/1925/gi_body.php). . Retrieved November 16, 2006.

[41] " Average Weather for Cairo, IL (http://www.weather.com/outlook/travel/businesstraveler/wxclimatology/monthly/graph/USIL0168)",weather.com

[42] " Chicago Weather (http://www.ustravelweather.com/weather-illinois/chicago-weather.asp)", ustravelweather.com

[43] " (http://www.weather.com/outlook/travel/businesstraveler/wxclimatology/monthly/graph/62025?from=36hr_bottomnav_business)",weather.com

[44] " Moline Weather (http://www.ustravelweather.com/weather-illinois/moline-weather.asp)", ustravelweather.com

[45] " Peoria Weather (http://www.ustravelweather.com/weather-illinois/peoria-weather.asp)", ustravelweather.com

[46] " Rockford Weather (http://www.ustravelweather.com/weather-illinois/rockford-weather.asp)", ustravelweather.com

[47] " Springfield Weather (http://www.ustravelweather.com/weather-illinois/springfield-weather.asp)", ustravelweather.com

[48] "Illinois Selected Social Characteristics in the United States: 2007" (http://factfinder.census.gov/servlet/ADPTable?_bm=y&-qr_name=ACS_2007_1YR_G00_DP2&-geo_id=04000US17&-ds_name=ACS_2007_1YR_G00_&-_lang=en&-redoLog=false). *2007 American Community Survey 1-Year Estimates*. U.S. Census Bureau. 2007. . Retrieved April 9, 2009.

[49] "Illinois QuickFacts" (http://quickfacts.census.gov/qfd/states/17000.html). U.S. Census Bureau. February 20, 2009. . Retrieved April 9, 2009.

[50] "Population and Population Centroid by State: 2000" (http://www.acsm.net/statecenters.html). American Congress on Surveying & Mapping. 2008. . Retrieved April 9, 2009.

[51] "Table 5. Estimates of Population Change for Metropolitan Statistical Areas and Rankings: July 1, 2007 to July 1, 2008" (http://hawaii.gov/dbedt/info/census/popestimate/2008_MSA_Hawaii/CBSA_EST2008_05.pdf) (PDF). *2008 Population Estimates*. U.S. Census Bureau. March 19, 2009. . Retrieved October 11, 2009.

[52] "Table 1: Annual Estimates of the Resident Population for Incorporated Places Over 100,000, Ranked by July 1, 2008 Population: April 1, 2000 to July 1, 2008 (SUB-EST2008-01)" (http://www.census.gov/popest/cities/tables/SUB-EST2008-01.xls). *2008 Population Estimates*. Population Division, United States Census Bureau. July 1, 2009. . Retrieved July 3, 2009.

[53] See Statemaster (http://www.statemaster.com/red/graph/peo_rom_cat_per_of_cat-people-roman-catholicism-percentage-catholics&int=-1&id=IL). Retrieved July 29, 2007.

[54] "The Association of Religion Data Archives | Maps & Reports" (http://www.thearda.com/mapsReports/reports/state/17_2000.asp). Thearda.com. . Retrieved September 14, 2009.

[55] "Newsroom – The Church of Jesus Christ of Latter-day Saints" (http://newsroom.lds.org/ldsnewsroom/eng/contact-us/usa-illinois). Newsroom.lds.org. . Retrieved February 7, 2011.

[56] "GDP by State" (http://greyhill.com/gdp-by-state). Greyhill Advisors. . Retrieved September 16, 2011.

[57] "Table 2. Annual Personal Income and Per Capita Personal Income by State and Region" (http://www.bea.gov/scb/pdf/2010/04 April/0410_SPI_tables.pdf). *Survey of Current Business – Bureau of Economic Analysis*. U.S. Department of Commerce. April 2010. . Retrieved April 24, 2010.

[58] "Current Unemployment Rates for States and Historical Highs/Lows" (http://www.bls.gov/web/laus/lauhsthl.htm). *Local Area Unemployment Statistics Information and Analysis*. U.S. Bureau of Labor Statistics. April 16, 2010. . Retrieved April 24, 2010.

[59] "Local Area Unemployment Statistics" (http://www.bls.gov/web/laus/laumstch.htm). Bureau of Labor Statistics. . Retrieved September 16, 2011.

[60] Pierog, Karen (January 12, 2011). "Illinois lawmakers pass big tax hike to aid budget" (http://www.reuters.com/article/2011/01/12/us-illinois-budget-idUSTRE70A6GP20110112). Reuters. . Retrieved February 7, 2011.

[61] Illinois Department of Revenue. Individual Income Tax (http://www.revenue.state.il.us/Businesses/TaxInformation/Income/individual.htm). Retrieved January 30, 2011.

[62] Illinois Department of Revenue. Illinois Sales Tax Reference Manual (PDF) (http://www.revenue.state.il.us/Publications/Sales/strrm/04012008/ST-25.pdf). p133. January 1, 2006.

[63] "Soybean Production by State 2008" (http://www.soystats.com/2009/Default-frames.htm). *Soy Stats*. The American Soybean Association. 2009. . Retrieved January 19, 2010.

[64] "Ethanol Fact Sheet" (http://www.ilcorn.org/internal.php?q=vprofile&id=90&date=&banner=ethanol). Illinois Corn Growers Association. 2010. . Retrieved January 18, 2010.

[65] "Manufacturing in Illinois" (http://www.commerce.state.il.us/NR/rdonlyres/1357C591-2810-4228-A334-E8B55EF1288D/0/ Manufacturing.pdf). Illinois Department of Commerce and Economic Opportunity. 2009. . Retrieved January 19, 2010.

[66] " Illinois in the Global Energy Marketplace (http://www.isgs.uiuc.edu/maps-data-pub/publications/energy01/globalm.shtml)", Robert Finley, 2001. Illinois State Geological Survey publication.

[67] Illinois State Geological Survey (http://www.isgs.uiuc.edu/). Coal in Illinois (http://www.isgs.illinois.edu/research/coal/ illinois-coal.shtml). Retrieved December 4, 2008.

[68] United States Department of Energy. Petroleum Profile: Illinois (http://tonto.eia.doe.gov/oog/info/state/il.html). Retrieved April 4, 2006.

[69] "Illinois Nuclear Industry" (http://www.eia.doe.gov/cneaf/nuclear/page/at_a_glance/states/statesil.html). U.S. Energy Information Administration. November 6, 2009. . Retrieved January 29, 2010.

[70] "Illinois Wind." Illinois Institute for Rural Affairs, Western Illinois University Illinoiswind.com (http://www.illinoiswind.org/index.asp)

[71] "Illinois Wind Activities" (http://www.windpoweringamerica.gov/astate_template.asp?stateab=il). *EERE*. U.S. Department of Energy. October 20, 2009. . Retrieved January 14, 2010.

[72] "U.S. Wind Energy Projects – Illinois" (http://www.awea.org/projects/Projects.aspx?s=Illinois). American Wind Energy Association. September 30, 2009. . Retrieved January 14, 2010.

[73] "Wind Power on the Illinois Horizon" (http://environmentalalmanac.blogspot.com/2006/09/wind-power-on-illinois-horizon.html), Rob Kanter, September 14, 2006. University of Illinois Environmental Council.

[74] "Illinois Renewable Electricity Profile" (http://www.eia.doe.gov/cneaf/solar.renewables/page/state_profiles/illinois.html). U.S. Energy Information Administration. 2007. . Retrieved January 15, 2010.

[75] Olbert, Lori (December 13, 2007). "Wind Farm Conference Tackles Complicated Issue" (http://centralillinoisproud.com/content/fulltext/ ?cid=5420). *CIProud.com*. WYZZ-TV/WMBD-TV. . Retrieved January 15, 2010.

[76] " BP Pledges $500 Million for Energy Biosciences Institute and Plans New Business to Exploit Research (http://www.bp.com/ genericarticle.do?categoryId=2012968&contentId=7018719)", BP.com, June 14, 2006.

[77] " Gov. Blagojevich joins Gov. Schwarzenegger, top BP executives to celebrate launch of $500 million biosciences energy research partnership with University of Illinois Urbana-Champaign, UC-Berkeley (http://www.illinois.gov/PressReleases/ShowPressRelease. cfm?SubjectID=2&RecNum=5690)". Press release, Illinois.gov. February 1, 2007.

[78] Mark McGuire Commentary (June 12, 2010). "Long look at Top 10 title droughts" (http://www.timesunion.com/news/article/ Long-look-at-Top-10-title-droughts-558915.php). Times Union. . Retrieved February 7, 2011.

[79] "The Longest Running Title Droughts in Sports" (http://bleacherreport.com/articles/ 404001-longest-playoff-droughts-in-the-major-sports). Bleacher Report. June 10, 2010. . Retrieved February 7, 2011.

[80] "Illinois & Michigan Canal" (http://www.nps.gov/ilmi). National Park Service. . Retrieved July 15, 2008.

[81] "Illinois" (http://www.nps.gov/state/il). National Park Service. . Retrieved July 15, 2008.

[82] S. Spacek, 2011 American State Litter Scorecard: New Rankings for an Increasingly Environmentallly Concerned Populous

[83] http://www.ioc.state.il.us/index.cfm/linkservid/468F0738-1CC1-DE6E-2F480FBBA9DB7BBA/showMeta/0/

[84] http://www.allthingspolitical.org/special_districts/illinois_library.htm

[85] http://www.lincolntrail.info/consultresources/Librarydistricthandbook2010.pdf

[86] http://woodstockpubliclibrary.org/content/rural-woodstock-public-library-district

[87] http://www.ilwastewater.org/memberlinks.htm

[88] "Census of State and Local Law Enforcement Agencies, 2000" (http://www.ojp.usdoj.gov/bjs/pub/ascii/csllea00.txt). *Bureau of Justice Statistics*. U.S. Department of Justice. October 2002. . Retrieved September 17, 2009.

[89] "Suburb shift turns state blue / The Christian Science Monitor" (http://www.csmonitor.com/2004/0716/p01s01-uspo.html). CSMonitor.com. July 16, 2004. . Retrieved February 7, 2011.

[90] "Chicgao's dominance puts Illinois solidly in 'blue-state' America. – Chicago Tribune (Chicago, IL)" (http://www.highbeam.com/doc/ 1G1-124280593.html). Highbeam.com. November 8, 2004. . Retrieved February 7, 2011.

[91] Pensoneau, Taylor (1997). *Governor Richard Ogilvie: in the interest of the state* (http://books.google.com/?id=bxNGKsylQXUC& pg=PA80&dq=kane+dupage+lake+cook+democratic&q=kane dupage lake cook democratic). Southern Illinois University Press. p. 314. ISBN 978-0809321483. . Retrieved September 23, 2009.

[92] Gimpel, James G.; Jason E. Schuknecht (2004). *Patchwork Nation: Sectionalism and Political Change in American Politics* (http://books. google.com/?id=rxQXjjzwrzEC&pg=PA359&dq="collar+counties"+hispanic+democratic#v=onepage&q="collar counties" hispanic democratic). University of Michigan Press. p. 488. ISBN 978-0472030309. . Retrieved September 23, 2009.

[93] "Illinois still blue despite Republican gains" (http://thecommunityword.com/online/blog/2010/12/08/ illinois-still-blue-despite-republican-gains/). The Community Word. December 8, 2010. . Retrieved February 7, 2011.

[94] "Illinois still a blue state ... kind of" (http://thesouthern.com/news/opinion/editorial/erickson/ article_22bcbc36-e9fe-11df-97e9-001cc4c03286.html). Thesouthern.com. November 7, 2010. . Retrieved February 7, 2011.

[95] Merriner, James L. (2004). *Grafters and Goo Goos: corruption and reform in Chicago, 1833–2003*. Carbondale: Southern Illinois University Press. ISBN 9780809325719. OCLC 52720998.

[96] "U.S. Senate: Art & History Home" (http://www.senate.gov/pagelayout/history/h_multi_sections_and_teasers/ Photo_Exhibit_African_American_Senators.htm). Senate.gov. . Retrieved February 7, 2011.

[97] "Best Colleges 2010 – National Universities Rankings" (http://colleges.usnews.rankingsandreviews.com/best-colleges/ national-universities-rankings/state+IL). *U.S. News & World Report*. August 19, 2009. . Retrieved March 23, 2010.

[98] "O'Hare International Airport Activity Statistics" (http://www.flychicago.com/Statistics/stats/1208ORDSUMMARY-REVISED.pdf). *FlyChicago.com*. City of Chicago. March 27, 2009. . Retrieved April 10, 2009.

[99] "Midway Airport Activity Statistics" (http://www.flychicago.com/Statistics/stats/1208SUMMARYRevised.pdf). *FlyChicago.com*. City of Chicago. January 30, 2009. . Retrieved April 10, 2009.

Further reading

- Bridges, Roger D.; Davis, Rodney O. (1984). *Illinois: its history & legacy*. St. Louis: River City Publishers. ISBN 0933150865. OCLC 11814096.

- Cole, Arthur Charles (1987) [1919]. *The era of the Civil War, 1848–1870*. Urbana: University of Illinois Press. ISBN 9780252013393. OCLC 14130434.

- Davis, James E. (1998). *Frontier Illinois*. Bloomington: Indiana University Press. ISBN 0-253-33423-3. OCLC 39182546.

- Gove, Samuel Kimball; Nowlan, James Dunlap (1996). *Illinois politics & government: the expanding metropolitan frontier*. Lincoln: University of Nebraska Press. ISBN 0-8032-7014-3. OCLC 33407256.

- Grossman, James R.; Keating, Ann Durkin; Reiff, Janice L. (2005) [2004]. *Electronic Encyclopedia of Chicago* (http://www.encyclopedia.chicagohistory.org/) (Online ed.). Chicago: Chicago Historical Society, Newberry Library. ISBN 0-226-31015-9. OCLC 60342627. Retrieved January 28, 2009.

- Hallwas, John E., ed (1986). *Illinois literature: the nineteenth century*. Macomb: Illinois Heritage Press. OCLC 14228886.

- Howard, Robert P. (1972). *Illinois; a history of the Prairie State*. Grand Rapids: W. B. Eerdmans Pub. Co. ISBN 0-8028-7025-2. OCLC 495362.

- Jensen, Richard E. (2001). *Illinois: a history*. Urbana: University of Illinois Press. ISBN 978-0-252-07021-1. OCLC 46769728.

- Keiser, John H. (1977). *Building for the centuries: Illinois, 1865 to 1898*. Urbana: University of Illinois Press. ISBN 978-0-252-00617-3. OCLC 2798051.

- Kilduff, Dorrell; Pygman, C. H. (1962). *Illinois; History, government, geography*. Chicago: Follett. OCLC 5223888.

- Kleppner, Paul (1988). *Political atlas of Illinois*. DeKalb: Northern Illinois University Press. ISBN 978-0-87580-136-0. OCLC 16755435.

- Meyer, Douglas K. (2000). *Making the heartland quilt: a geographical history of settlement and migration in early-nineteenth-century Illinois* (http://www.questia.com/PM.qst?a=o&d=65659204). Carbondale: Southern Illinois University Press. ISBN 978-0-585-37905-0. OCLC 48139026.

- Nowlan, James D.; Gove, Samuel K.; Winkel, Richard J. (2010). *Illinois Politics: A Citizen's Guide*. Urbana: University of Illinois Press. ISBN 978-0252077029.

- Sutton, Robert P. (1976). *The Prairie State; a documentary history of Illinois*. Grand Rapids: Eerdmans. ISBN 0-8028-1651-7. OCLC 2603998.

- Walton, Clyde C. (1970). *An Illinois reader*. DeKalb: Northern Illinois University Press. ISBN 978-0-87580-014-1. OCLC 89905.

- Works Progress Administration (1983) [1939]. *The WPA guide to Illinois: the Federal Writers' Project guide to 1930s Illinois*. New York: Pantheon Books. ISBN 978-0-394-72195-8. OCLC 239788752.

External links

* State of Illinois (http://www.illinois.gov) *official government website*
* The Official Website for International Visitors to Chicago (http://www.gochicago.com/)
* Illinois (http://www.dmoz.org/Regional/North_America/United_States/Illinois/) at the Open Directory Project
* State of Illinois (http://www.library.illinois.edu/doc/collections/stateofillinois.html) research information guide from the University of Illinois at Urbana–Champaign
* Illinois Bureau of Tourism (http://www.enjoyillinois.com)
* Illinois: Science In Your Backyard (http://www.usgs.gov/state/state.asp?State=IL) – USGS
* Illinois State Agency Databases (http://wikis.ala.org/godort/index.php/Illinois) – compiled by the Government Documents Round Table (GODORT) of the American Library Association
* Illinois State Energy Profile (http://tonto.eia.doe.gov/state/state_energy_profiles.cfm?sid=IL) – DOE, Energy Information Administration
* Illinois: State Fact Sheets (http://www.ers.usda.gov/StateFacts/Il.htm) – USDA, Economic Research Service
* Illinois State Guide (http://www.loc.gov/rr/program/bib/states/illinois//) – LOC, Virtual Programs & Services
* Biographies Of Governors of Illinois: 1818 to 1885 (http://www.onlinebiographies.info/gov/il/index.htm) – compiled by OnlineBiographies.info
* Illinois Highways Page (http://www.n9jig.com) – by Richard Carlson
* OpenStreetMap has geographic data related to Illinois (http://www.openstreetmap.org/browse/relation/122586)

gag:İllinois mrj:Иллинойс (штат)

Article Sources and Contributors

Illinois_Route_105 *Source*: http://en.wikipedia.org/w/index.php?title=Illinois_Route_105 *Contributors*: Bigturtle, Black Falcon, Daniel73480, Detcin, Hadrianheugh, Imzadi1979, Lpangelrob, Nickvet419, Omnedon, Polaron, Sarjeant, WOSlinker, 4 anonymous edits

Illinois_Department_of_Transportation *Source*: http://en.wikipedia.org/w/index.php?title=Illinois_Department_of_Transportation *Contributors*: 978221sf, Adrianadams62896, Alansohn, Arpingstone, Astuishin, Bigturtle, Buphitau, Ce6, Fitzaubrey, Funandtrvl, Hmains, Itbzach, JohnnyAlbert10, Master son, Microcell, Miller17CU94, Nudecline, Overpush, Presidentman, Rrius, Shawn in Montreal, Smalljim, Spiffy sperry, TwinsMetsFan, WhisperToMe, 31 anonymous edits

Illinois_Route_48 *Source*: http://en.wikipedia.org/w/index.php?title=Illinois_Route_48 *Contributors*: Black Falcon, Detcin, Freakofnurture, Imzadi1979, Jaranda, Lpangelrob, Nickvet419, Omnedon, Polaron, Rschen7754, WOSlinker, 8 anonymous edits

Interstate_72 *Source*: http://en.wikipedia.org/w/index.php?title=Interstate_72 *Contributors*: Albinfo, Artisol2345, Aspects, Bkell, Bluemoose, Bobblewik, Bryan Derksen, Bumm13, C.Fred, Catbar, Cmdrjameson, Davidmac2003, Denelson83, Discordanian, Dough4872, Dual Freq, Fredddie, Freewayguy, Gpietsch, Ground Zero, Hamjamguy, Hmains, I-10, JAldrich73, JohnnyAlbert10, Jpers36, Lazytiger, Lpangelrob, Master son, Mhking, Michael Patrick, Morriswa, NE2, Nk, Onagadori, Ownage2214, Patricknoddy, Paul from Michigan, Plastikspork, Polaron, Potatoswatter, Robertb-dc, Rschen7754, Rt66lt, SPUI, Sarjeant, Sfoskett, Skitters, Southern Illinois SKYWARN, Stratosphere, Synchronism, TwinsMetsFan, VerruckteDan, WOSlinker, WillC, Xenon54, 72 anonymous edits

Bement,_Illinois *Source*: http://en.wikipedia.org/w/index.php?title=Bement%2C_Illinois *Contributors*: Antandrus, Ascidian, Ballstothewalls12345, Bevo74, Bigconnor30, CWii, Coronaboy, Docu, Dual Freq, Escape Orbit, Fastily, Iridescent, Mickeym191, Pearle, Quartermaster, RAMILES, Ram-Man, Ridinlow333, Saluki2002, Sgt Pinback, TheCatalyst31, Tremont798, Unschool, Wachholder0, 61 anonymous edits

Illinois_Route_104 *Source*: http://en.wikipedia.org/w/index.php?title=Illinois_Route_104 *Contributors*: Bigturtle, Black Falcon, Detcin, Docu, Freakofnurture, Haffy54, Imzadi1979, Lpangelrob, MrRadioGuy, NE2, Nickvet419, Omnedon, Polaron, WOSlinker, 6 anonymous edits

Illinois_Route_47 *Source*: http://en.wikipedia.org/w/index.php?title=Illinois_Route_47 *Contributors*: Alai, BL Lacertae, Black Falcon, Ce6, Detcin, Drp1103, Ebyabe, Freakofnurture, Fredddie, Geognerd, Goethean, Gws57, Hairmetal1990, Imzadi1979, Lpangelrob, NBA2020, NE2, Nickvet419, Omnedon, Polaron, Rschen7754, SPUI, Sarjeant, Sortior, Speciate, Stoneburner132, That Guy, From That Show!, TheCatalyst31, Verloren813, WOSlinker, WikiDon, 19 anonymous edits

Macon_County,_Illinois *Source*: http://en.wikipedia.org/w/index.php?title=Macon_County%2C_Illinois *Contributors*: AshyLarry, Betacommand, CanisRufus, D6, Danielba894, Danny, Dual Freq, Erieillinois, Factitious, Fishal, Frenkmelk, GrahamHardy, Hephaestos, Hu12, Johnpacklambert, Kranar drogin, M dorothy, Monegasque, NewAgeParanormalResearch, Notheotherhorizon, Omnedon, Ram-Man, Robertjohnsonrj, Rrius, Sgt Pinback, Smallbones, Sortior, Spiffy sperry, Template namespace initialisation script, Tim!, Tom harrison, Ulric1313, Utopies, Voyager, Whhalbert, WillC, Worldenc, 13 anonymous edits

Piatt_County,_Illinois *Source*: http://en.wikipedia.org/w/index.php?title=Piatt_County%2C_Illinois *Contributors*: Acntx, CanisRufus, D6, Danny, Dual Freq, Fishal, Fogie2003, GrahamHardy, Hephaestos, Joseph Solis in Australia, Kkemper, M dorothy, Monegasque, NERIUM, Omnedon, PeaceFrog201, Peter200, Ram-Man, SangamonTownship, Sgt Pinback, Smallbones, Sortior, Spiffy sperry, Template namespace initialisation script, Tim!, TwinCityIL, Utopies, Voyager, Whhalbert, WillC, Worldenc, 8 anonymous edits

Bryant_Cottage_State_Historic_Site *Source*: http://en.wikipedia.org/w/index.php?title=Bryant_Cottage_State_Historic_Site *Contributors*: Belovedfreak, Bigturtle, Bill Thayer, BrownHairedGirl, Cbustapeck, Chris the speller, Deor, Deville, Dmadeo, Gardar Rurak, IvoShandor, Jllm06, Rich257, Vegaswikian1, 1 anonymous edits

Illinois *Source*: http://en.wikipedia.org/w/index.php?title=Illinois *Contributors*: -ross616-, 0, 0v1d1o114, 16@r, 1916b, 21655, 2help, 55david, 65.197.2.xxx, 7&6=thirteen, A.J.A., A.villarini, A1017762, A2Kafir, AKeen, AOC25, AP1787, ARCHSupr, Aaron Brenneman, Aaron Walden, AaronCBurke, Abc518, Abductive, Abog, Aboutmovies, Academic Challenger, Access Denied, Acjelen, Acroterion, AdamNewcomb, Adambro, Addshore, AdjustShift, AdultSwim, Aeusoes1, After Midnight, Agriculture, Ahassan05, Ahoerstemeier, Ahuskay, Airconswitch, Aitias, Aivazovsky, Alansmithee42, Alansohn, Aleenf1, AlexiusHoratius, Alexwcovington, Allen4names, Allstarecho, Alphachimp, Altenmann, Amazonien, Amikeco, Andrewpmk, Andrewprest, Andy Marchbanks, Andy120290, Anetode, Angech01, AngelOfSadness, Angr, Angrysockhop, Animum, Antandrus, ApplesMe, Appraiser, Arakunem, Arch dude, Arda Xi, Ardengoldstein, Arjayay, Artemisboy, Arthena, Ashford11, Astuishin, AuburnPilot, Audra4, Avb, Avenged Eightfold, Avram, Aztom2, BSveen, Backslash Forwardslash, Backyard Wiffleball, Bagatelle, Balcer, Bardocksaian, Barek, BaronLarf, BartBenjamin, Basel kamoua, Batteriesnotincluded, Bbballa94, Bdiscoe, Beanie315, Beatgr, Beetstra, Before My Ken, Beland, Belgrano, BenBaker, Bencherlite, Bender235, Benjh40, Bentley4, Berean Hunter, Betta18, Beygle, Bfigura's puppy, Bgtyui, Bhuck, Big iron, Big juicy breast, BigBang11, Bignd500, Bigwood21, Billwhittaker, Binary TSO, Bjszp9, Bkonrad, Bkopale, Blades123546, Blake1445, Blanchardb, BlueAzure, Bnm1212, Boat boat b0at, Bobo192, Bobtheblobrocks, Boing! said Zebedee, Bomac, Bonadea, Bongwarrior, Boomshadow, BorgHunter, BradBeattie, Bradl333, Brainboy109, Branddobbe, Brandonh, BrendelSignature, Breno, Brianboru, Bridgecross, Brion VIBBER, Brother Officer, Bsadowski1, Buaidh, Bunnyhop11, Burgwerworldz, Burntsauce, Burzmali, Buster7, Bwrs, C.Logan, CJLL Wright, CJLippert, CSWarren, CWii, Cableshaft, Caltas, CambridgeBayWeather, Can't sleep, clown will eat me, Canadian Bobby, CanadianLinuxUser, Candyhyewon, CanisRufus, Canseb, Caponer, Capricorn42, Captain Caveman, CardinalDan, Carlaude, Carlw4514, Catgut, Cburnett, Cchow2, Ceezmad, Ceyockey, Cfrdc, Cftgb, Charles Smith 1000, CharlotteWebb, Chicago god, Chicago103, Chill doubt, ChongDae, Choster, Chris the speller, ChrisIrk02, Chrisp18, Chun-hian, Ciaccona, Civil Engineer III, Ckatz, ClairSamoht, ClamDip, ClockworkLunch, Cmflores1, Cmprince, Cnilep, Colonies Chris, Computerjoe, Conversion script, CoolKid1993, Coolcaesar, Coolio130, Corpx, Corriebertus, Cosmo1976, Countercouper, Courcelles, Cpastern, Craig2434, CrazyC83, Creez34, Crimson3981, Cubs197, CuentaDisponible, Curieux, Cvbgf, CylonCAG, D, D6, DARTH SIDIOUS 2, DCEdwards1966, DDCrunch, DEUCE, DLJessup, DS1953, Dabomb87, Dale Arnett, DanielCD, Danno uk, Danny, DannyQuack, Danpenning, Dark567, Darth Mike, Daven200520, David Fell, DavidWBrooks, Dawnseeker2000, Dblood93, Dcornwall, Deadcatmuseum, Deconstructhis, Deflective, Delirium, Delldot, Dellpie, Deltabeignet, Demonic Vampire Lord, Denisarona, Deor, DerHexer, Derontae23, Destructicus, DevinCook, Dewelar, Dgrade2, Dharmesh255, Diannaa, Dimadick, Discospinster, Dismas, DivineIntervention, Djcosta, Dkdantastic, Does anybody love sex, Does anybody want to have sex, Doniago, Dori, Dp462090, Dr. Blofeld, Dragon.zuhair, Dralwik, Drumlineramos, Dsnrabble, Dual Freq, Dudester99, Dwilso, Dysepsion, Dysprosia, ESkog, Eco84, Ed, Ed g2s, Edivorce, Efvgu, Elite18, Elkman, Elphion, Emeraldcityserendipity, EncycloPetey, Eouw0o83hf, Epbr123, Epolk, Ericnotderek, Erub, EscapingLife, Etams, EurekaLott, Evdog2k3, Evercat, Evice, Evlekis, Ewulp, Excirial, FF2010, FJPB, Fabalous Sex, Favonian, Feinoha, Fepaz0024, Fishal, Fishing, Fishy135, Flatterworld, Floquenbeam, Flowerpotman, FluffyDog13, Flyguy649, Fogherty V. Tatin, Fraggle81, Freakmighty, Freakofnurture, FredR, Frenkmelk, FreplySpang, Froghermet, Frosted14, Frymaster, Funandtrvl, Funnyhat, Gail, Gaius Cornelius, Garagepunk, Garkeith, Gary Cziko, Gdo01, Gene Nygaard, Geologyguy, Gerald Farinas, Ghiraddje, Gilliam, Gimmetrow, Glane23, Glidepath07k, GoPurpleNGold24, Godlvall2, Goethean, Gogo Dodo, GoldenGlory84, Good Olfactory, GoodDay, Gpyoung, GraemeMcRae, Grafen, Graham87, Grandexandi, GregU, Grey Wanderer, Greybeard, Grim23, Gscshoyru, Gujuguy, Gurch, Guug13, Gwernol, Gwhose, Gwmayes, Hadal, Hager jeff, HalfShadow, Hamako, Hamgoblin45, Hansonmi, HappyCamper, Hatmatbbat10, Hdt83, Hequals2henry, Higbvuyb, Highfields, Hippymac, Hmains, Hoary, Hockeyjake4, Hoejoehoe, HollyAm, Hoosierdaddy2, Horologium, Howcheng, Howdybob, Hudson Stern, Hushpuckena, HuskyHuskie, Husond, Hydrogen Iodide, IMatthew, IRelayer, Iadrian yu, Ijmki, Ijnhy, Iketsi, Ilandedonmyhead, Ilikegrls2043, Illinois1, Immunize, Indon, Ineky, Infidel taco, Instinct, Iohannes Animosus, Iridescent, Irishguy, Iterator12n, Ittod, J.delanoy, JCDenton2052, JForget, JNW, JPFay, JW1805, Jackfork, Jackol, Jacksonbs, Jacob.jose, Jagislaqroo, Jakubz, Jarry1250, Jatkins, Javert, Jaxl, Jbamb, Jcace91, Jcam, Jcrocker, Jeditor17, Jeff Wheeler, Jeffko, Jeffkw, Jengod, Jennymodglin, Jessten95, Jhendin, JimIrwin, Jimmuldrow, Jimmy Pitt, JinJian, Jivecat, Jketola, Jmlk17, JoJoVader12, JoeSmack, Joffeloff, Johaen, John K, John of Reading, JohnDBuell, Johnsemlak, Jojhutton, Jonathan97X, JosephNorton, Josiahatwood, Jpers36, Jtrost, Juke2, Juliancolton, Jultemplet, Jumpcho, Jusdafax, Juzeris, K1Bond007, KUsam, Ka Faraq Gatri, Kablammo, Kahuroa, Kalathalan, Kamezuki, Kaneland22, Kansan, Kapal345, Karam.Anthony.K, Kbdank71, Keelan2000, Kelly Martin, Ken g6, Khatru2, Kilo-Lima, Kilon22, King of Hearts, Kingpin13, Kintetsubuffalo, Kirrages, KnightLago, Knotwork, KnowledgeOfSelf, Knowledgefreak100, Kontar, Kookie131, Kosjeyr, Koug, Koupa, Kozuch, Kralizec!, Kranar drogin, Kristinpedia, Kristof vt, Kseferovic, Kuru, Kwamikagami, KyuuA4, L Glidewell, L Kensington, L L Weiskirch, Lala6, Laurascudder, LaurenLou11, LeVoyageur, Leoni2, Leuko, Levineps, Lexi Marie, Lfh, Libertyville, Lightdarkness, Lightmouse, Lights, Lilac Soul, Lilly15, LindsayH, Little Mountain 5, Ljmajer, Lmbstl, Locquincer, Logan, Lolimbored, LongBay, Lookatall, Loompyloompy313, Looxix, Lord Voldemort, Lpangelrob, LtNOWIS, Luna Santin, M C Y 1008, M dorothy, MJCdetroit, MJDTed, MK8, MacTire02, Madcoverboy, Maelnuneb, Magioladitis, Magister Mathematicae, Making observations, Malathos, Malik Shabazz, Manicsfan, Manuel Trujillo Berges, Marcusmax, Marcusscotus1, Mareino, Marek69, Martarius, Masterofpuppets792, Masterpiece2000, Matijap, Matt Yeager, Mattgirling, Mattjordan4, Mattle20, Mav, Maxy93, Mcshadypl, Mdawwg, Mdebets, Meelar, Meisterkoch, Mentifisto, Metagraph, Metro Sex, Metropolitan90, Mgeheren, Mhghu, Mhking, Michael Hardy, Michaeldsuarez, MickWest, Midnight Green, Mifter, Miguel Andrade, Mike s, Mikeae, Mikeknein, Mikeo, Mikevegas40, Minesweeper, Minimeb6, Mirth, MisfitToys, Miskwito, Mistercow, Mwilso24, Mxn, My toof is broke, MyReference, Myfriendbrenn, NEICenergy, NEVETS04, Nakon, Nancy, Natgeoguy, NatusRoma, NawlinWiki, Nczcar01, Neko-chan, NellieBly, Nephron, Nestorius, Neutrality, Neverquick, Nibuod, Niels0827, Nightfox939, Njuk, Nkayesmith, Node ue, Nonnospizza, Noozgroop, NorCalHistory, North Shoreman, NorwegianBlue, Novis-M, Nsaa, Nubiatech, Nunh-huh, Nyttend, Oahfapgah, Oda Mari, Ogdeniam, Ohconfucius, Ohnoitsjamie, Oliver202, Omeomi, OneNester, Onixz100, Oo64eva, Optikos, Orangemike, Orestek, Orlady, Ouatic-7, Owen, Oxymoron83, PL290, Pakibrat, Parkwells, Patrick, PatrickAce, Patrickneil, Patstuart, Paul August, Paul H., Paxse, Paxsimius, Pekinpekin, Pepper, Persian Poet Gal, Peter Karlsen, Petersjh72, Petrb, Petronas, Pfly, Philip Trueman, Phillip J, Phxsuns562, PigFlu Oink, Pigman, Pinethicket, Pleasant Peter, Poeloq, Pokemonblackds, PostdIf, Pozcircuitboy, Prashanthns, Prchristensen, Qaqaq, Qqqqqq, QuartierLatin1968, Queenqpawn, Qxz, R'n'B, R.T.Gellar, R4r5t6, RJaguar3, RL0919, RTC, Racepacket, RadioFan, Raedwulf16, Rande M Sefowt, Randyc, RattusMaximus, Ratwod, Ravedave, Raven4x4x, Ray13, Raymond Meredith, Rdcfr, Rebelwithnocause, Red Director, Redmarkviolinist, Redslippers, Reed Jones, Retro Sex, Revas, RexNL, Reyellogreeblu, Rhatsa26X, Rhino6187, Rhrad, Rich Farmbrough, Richtom80, Rick Block, RickK, Rjanag, Rjensen, Rjwilmsi, Rklawton, Rmhermen, Rob Shepard, Robocoder, Robofish,

Image Sources, Licenses and Contributors

File:Illinois 105.svg *Source*: http://en.wikipedia.org/w/index.php?title=File:Illinois_105.svg *License*: unknown *Contributors*: SPUI

Image:Illinois 48.svg *Source*: http://en.wikipedia.org/w/index.php?title=File:Illinois_48.svg *License*: unknown *Contributors*: SPUI

Image:I-72.svg *Source*: http://en.wikipedia.org/w/index.php?title=File:I-72.svg *License*: unknown *Contributors*: Augiasstallputzer, Ltljltlj, SPUI, 1 anonymous edits

file:IL DOT.svg *Source*: http://en.wikipedia.org/w/index.php?title=File:IL_DOT.svg *License*: unknown *Contributors*: Overpush

File:IDOT districts.png *Source*: http://en.wikipedia.org/w/index.php?title=File:IDOT_districts.png *License*: unknown *Contributors*: User:Dual Freq, User:Presidentman

File:Illinois 48.svg *Source*: http://en.wikipedia.org/w/index.php?title=File:Illinois_48.svg *License*: unknown *Contributors*: SPUI

Image:I-55.svg *Source*: http://en.wikipedia.org/w/index.php?title=File:I-55.svg *License*: unknown *Contributors*: Augiasstallputzer, Fran Rogers, Ltljltlj, SPUI, Xnatedawgx, 1 anonymous edits

Image:Illinois 127.svg *Source*: http://en.wikipedia.org/w/index.php?title=File:Illinois_127.svg *License*: unknown *Contributors*: Common Good, SPUI

Image:Illinois 54.svg *Source*: http://en.wikipedia.org/w/index.php?title=File:Illinois_54.svg *License*: unknown *Contributors*: Closeapple, SPUI

Image:Dewitt County Route 2 IL.svg *Source*: http://en.wikipedia.org/w/index.php?title=File:Dewitt_County_Route_2_IL.svg *License*: unknown *Contributors*: Ltljltlj

File:I-72.svg *Source*: http://en.wikipedia.org/w/index.php?title=File:I-72.svg *License*: unknown *Contributors*: Augiasstallputzer, Ltljltlj, SPUI, 1 anonymous edits

File:Interstate 72 map.png *Source*: http://en.wikipedia.org/w/index.php?title=File:Interstate_72_map.png *License*: unknown *Contributors*: w:User:StratosphereNick Nolte

File:No image wide.svg *Source*: http://en.wikipedia.org/w/index.php?title=File:No_image_wide.svg *License*: unknown *Contributors*: SPUI, TwinsMetsFan

File:Business plate.svg *Source*: http://en.wikipedia.org/w/index.php?title=File:Business_plate.svg *License*: unknown *Contributors*: Homefryes, Ltljltlj, RTCNCA, SPUI

File:US 36.svg *Source*: http://en.wikipedia.org/w/index.php?title=File:US_36.svg *License*: unknown *Contributors*: Bidgee, SPUI, Xnatedawgx, 3 anonymous edits

File:US 61.svg *Source*: http://en.wikipedia.org/w/index.php?title=File:US_61.svg *License*: unknown *Contributors*: Bidgee, SPUI, Xnatedawgx, 3 anonymous edits

File:I-172.svg *Source*: http://en.wikipedia.org/w/index.php?title=File:I-172.svg *License*: unknown *Contributors*: Common Good, Ltljltlj, SPUI, Xnatedawgx

File:US 54.svg *Source*: http://en.wikipedia.org/w/index.php?title=File:US_54.svg *License*: unknown *Contributors*: SPUI, Xnatedawgx, 1 anonymous edits

File:US 67.svg *Source*: http://en.wikipedia.org/w/index.php?title=File:US_67.svg *License*: unknown *Contributors*: Bidgee, SPUI, Xnatedawgx, 2 anonymous edits

File:I-55.svg *Source*: http://en.wikipedia.org/w/index.php?title=File:I-55.svg *License*: unknown *Contributors*: Augiasstallputzer, Fran Rogers, Ltljltlj, SPUI, Xnatedawgx, 1 anonymous edits

File:US 51.svg *Source*: http://en.wikipedia.org/w/index.php?title=File:US_51.svg *License*: unknown *Contributors*: Bidgee, SPUI, 3 anonymous edits

File:I-57.svg *Source*: http://en.wikipedia.org/w/index.php?title=File:I-57.svg *License*: unknown *Contributors*: Augiasstallputzer, Ltljltlj, Rfc1394, SPUI, Xnatedawgx

Image:I-72 North of Seymour Illinois.jpg *Source*: http://en.wikipedia.org/w/index.php?title=File:I-72_North_of_Seymour_Illinois.jpg *License*: unknown *Contributors*: User:Dual Freq

File:MO-79.svg *Source*: http://en.wikipedia.org/w/index.php?title=File:MO-79.svg *License*: unknown *Contributors*: User:Vishwin60

File:MO-supp-N.svg *Source*: http://en.wikipedia.org/w/index.php?title=File:MO-supp-N.svg *License*: unknown *Contributors*: User:PHenry

File:Illinois 106.svg *Source*: http://en.wikipedia.org/w/index.php?title=File:Illinois_106.svg *License*: unknown *Contributors*: SPUI

File:Illinois 96.svg *Source*: http://en.wikipedia.org/w/index.php?title=File:Illinois_96.svg *License*: unknown *Contributors*: SPUI

File:Illinois 107.svg *Source*: http://en.wikipedia.org/w/index.php?title=File:Illinois_107.svg *License*: unknown *Contributors*: SPUI

File:Illinois 100.svg *Source*: http://en.wikipedia.org/w/index.php?title=File:Illinois_100.svg *License*: unknown *Contributors*: Rocket000, SPUI

File:Business Loop 72.svg *Source*: http://en.wikipedia.org/w/index.php?title=File:Business_Loop_72.svg *License*: unknown *Contributors*: I-215, KelleyCook, Ltljltlj, Mountain169257, SPUI, T2, 1 anonymous edits

File:Illinois 267.svg *Source*: http://en.wikipedia.org/w/index.php?title=File:Illinois_267.svg *License*: unknown *Contributors*: Closeapple, SPUI

File:Illinois 104.svg *Source*: http://en.wikipedia.org/w/index.php?title=File:Illinois_104.svg *License*: unknown *Contributors*: SPUI

File:Illinois 123.svg *Source*: http://en.wikipedia.org/w/index.php?title=File:Illinois_123.svg *License*: unknown *Contributors*: SPUI

File:Illinois 4.svg *Source*: http://en.wikipedia.org/w/index.php?title=File:Illinois_4.svg *License*: unknown *Contributors*: SPUI

File:Business Loop 55.svg *Source*: http://en.wikipedia.org/w/index.php?title=File:Business_Loop_55.svg *License*: unknown *Contributors*: I-215, KelleyCook, Ltljltlj, Mountain169257, SPUI, T2, 1 anonymous edits

File:Illinois 29.svg *Source*: http://en.wikipedia.org/w/index.php?title=File:Illinois_29.svg *License*: unknown *Contributors*: Closeapple, SPUI

File:Illinois 97.svg *Source*: http://en.wikipedia.org/w/index.php?title=File:Illinois_97.svg *License*: unknown *Contributors*: Common Good, SPUI

File:Illinois 121.svg *Source*: http://en.wikipedia.org/w/index.php?title=File:Illinois_121.svg *License*: unknown *Contributors*: SPUI

File:Illinois 10.svg *Source*: http://en.wikipedia.org/w/index.php?title=File:Illinois_10.svg *License*: unknown *Contributors*: SPUI

File:Illinois 47.svg *Source*: http://en.wikipedia.org/w/index.php?title=File:Illinois_47.svg *License*: unknown *Contributors*: Closeapple, SPUI

File:I-74.svg *Source*: http://en.wikipedia.org/w/index.php?title=File:I-74.svg *License*: unknown *Contributors*: Augiasstallputzer, Ltljltlj, SPUI, Xnatedawgx, 1 anonymous edits

Image: Illinois Locator Map.PNG *Source*: http://en.wikipedia.org/w/index.php?title=File:Illinois_Locator_Map.PNG *License*: unknown *Contributors*: US Census, Ruhrfisch

File:Red pog.svg *Source*: http://en.wikipedia.org/w/index.php?title=File:Red_pog.svg *License*: unknown *Contributors*: Anomie

Image: Map of USA IL.svg *Source*: http://en.wikipedia.org/w/index.php?title=File:Map_of_USA_IL.svg *License*: unknown *Contributors*: Abnormaal, Hogweard, Huebi, Lokal Profil, Lupo, Mattbuck, Petr Dlouhý, 1 anonymous edits

Image:Bement Illinois High School.jpg *Source*: http://en.wikipedia.org/w/index.php?title=File:Bement_Illinois_High_School.jpg *License*: unknown *Contributors*: User:Dual Freq

Image:US 24.svg *Source*: http://en.wikipedia.org/w/index.php?title=File:US_24.svg *License*: unknown *Contributors*: Bidgee, SPUI, 3 anonymous edits

Image:Illinois 57.svg *Source*: http://en.wikipedia.org/w/index.php?title=File:Illinois_57.svg *License*: unknown *Contributors*: SPUI

Image:GreatRiverRoad.svg *Source*: http://en.wikipedia.org/w/index.php?title=File:GreatRiverRoad.svg *License*: unknown *Contributors*: User:Master_son

Image:Illinois 29.svg *Source*: http://en.wikipedia.org/w/index.php?title=File:Illinois_29.svg *License*: unknown *Contributors*: Closeapple, SPUI

Image:Illinois 10.svg *Source*: http://en.wikipedia.org/w/index.php?title=File:Illinois_10.svg *License*: unknown *Contributors*: SPUI

Image:WIS 120.svg *Source*: http://en.wikipedia.org/w/index.php?title=File:WIS_120.svg *License*: unknown *Contributors*: Juliancolton, SPUI

File:WIS 120.svg *Source*: http://en.wikipedia.org/w/index.php?title=File:WIS_120.svg *License*: unknown *Contributors*: Juliancolton, SPUI

File:Illinois 173.svg *Source*: http://en.wikipedia.org/w/index.php?title=File:Illinois_173.svg *License*: unknown *Contributors*: Common Good, SPUI

File:Illinois 120.svg *Source*: http://en.wikipedia.org/w/index.php?title=File:Illinois_120.svg *License*: unknown *Contributors*: SPUI

File:US 14.svg *Source*: http://en.wikipedia.org/w/index.php?title=File:US_14.svg *License*: unknown *Contributors*: Bidgee, SPUI, 3 anonymous edits

File:Illinois 176.svg *Source*: http://en.wikipedia.org/w/index.php?title=File:Illinois_176.svg *License*: unknown *Contributors*: Common Good, SPUI

File:I-90.svg *Source*: http://en.wikipedia.org/w/index.php?title=File:I-90.svg *License*: unknown *Contributors*: Augiasstallputzer, Ltljltlj, SPUI, 2 anonymous edits

File:US 20.svg *Source*: http://en.wikipedia.org/w/index.php?title=File:US_20.svg *License*: unknown *Contributors*: Kanonkas, SPUI

File:Illinois 72.svg *Source*: http://en.wikipedia.org/w/index.php?title=File:Illinois_72.svg *License*: unknown *Contributors*: Closeapple, SPUI

File:Illinois 64.svg *Source*: http://en.wikipedia.org/w/index.php?title=File:Illinois_64.svg *License*: unknown *Contributors*: Closeapple, SPUI

File:Illinois 38.svg *Source*: http://en.wikipedia.org/w/index.php?title=File:Illinois_38.svg *License*: unknown *Contributors*: SPUI

File:I-88.svg *Source*: http://en.wikipedia.org/w/index.php?title=File:I-88.svg *License*: unknown *Contributors*: Augiasstallputzer, Kanonkas, Ltljltlj, SPUI

File:US 30.svg *Source*: http://en.wikipedia.org/w/index.php?title=File:US_30.svg *License*: unknown *Contributors*: Bidgee, SPUI, 4 anonymous edits

File:Illinois 56.svg *Source*: http://en.wikipedia.org/w/index.php?title=File:Illinois_56.svg *License*: unknown *Contributors*: SPUI

File:US 34.svg *Source*: http://en.wikipedia.org/w/index.php?title=File:US_34.svg *License*: unknown *Contributors*: Bidgee, SPUI, Xnatedawgx, 3 anonymous edits

File:Illinois 126.svg *Source*: http://en.wikipedia.org/w/index.php?title=File:Illinois_126.svg *License*: unknown *Contributors*: SPUI

Printed by Books on Demand GmbH, Norderstedt / Germany